Excel

HSC & PRELIMINARY

DESIGN AND TECHNOLOGY

Get the Results You Want!

DEBORAH TREVALLION
&
RUSSELL TRIMMER

PASCAL
PRESS

Reprinted 2012
Sample HSC Examinations updated for new format 2015
Reprinted 2017, 2021, 2024

ISBN 978 1 74125 309 2

Pascal Press
PO Box 250
Glebe NSW 2037
(02) 9198 1748
www.pascalpress.com.au

Publisher: Vivienne Joannou
Project editor: Mark Dixon
Edited by Ian Rohr
Typeset by Grizzly Graphics (Leanne Richters)
Cover by Dizign Pty Ltd
Photos by iStock and Dreamstime
Printed by Vivar Printing/Green Giant Press

Acknowledgements
I thank my Design and Technology marking colleagues who have provided some of the most inspiring, selfless and productive experiences I have ever encountered. I thank my family—Trish, Matt and Lanni— for allowing me time off to explore my ideas. I thank my co-author, Deborah, who was so purposeful in refining the vision. My inspiration comes from Ernest Shackleton and his failed crossing of the South Pole: his only true failure would have been not giving it a go. (Russell)

I would like to thank my family, especially Talysa and Mitchel, for their kindness, love, patience and support; my friends for their time and understanding; and my co-author and colleagues for their hard work, inspirational ideas and for improving the way I think. Words to live by: let your life be an adventure, look for the invisible, listen to the silences, touch your imagination and make each day something special. (Deborah)

Students
All care has been taken in the preparation of this study guide, but please check with your teacher or the NSW Education Standards Authority about the exact requirements of the course you are studying as these can change from year to year.

The validity and appropriateness of the internet addresses (URLs) in this book were checked at the time of publication. Due to the dynamic nature of the internet, the publisher cannot accept responsibility for the continued validity or content of these web addresses.

Contents

Introduction

Design and Technology is a course that will inspire, fulfil and motivate. It involves designing, creating or synthesising, bringing out the best of human creativity, lateral thinking and innovative ideas as well as using the latest available technology to bring designs to life.

Today we use advanced technology to make everyday activities appear simple. It is important to gain an understanding of how these technologies work in order to create new technologies for both today and the future.

Design and Technology is about inventing practical solutions to problems through imaginative thinking. Students use a range of tools, materials, skills and techniques to research, design and manufacture products, systems and environments that meet needs. This course will allow you to work in a stimulating environment, engage in personal development and learn to manage and use technologies that will develop tomorrow's innovations.

Within this course you are required to complete a major design project worth sixty marks. This design project may be a product, system or environment. It must be supported by a technology-based design folio that communicates every step of the project's evolution. There is also the written exam that is worth forty per cent. These two marks are added together to form a final mark that is used, with your assessment marks, when your Higher School Certificate mark is calculated. For more details of assessment and syllabus procedures refer to the NESA website at:
www.educationstandards.nsw.edu.au

This study guide has been written to support your classroom learning and provide additional information that will extend your skills, knowledge and understanding. The following table links the chapters in the guide to the Design and Technology Stage 6 Syllabus Outcomes.

Syllabus Outcomes		Study guide chapters
P1.1	Examines design theory and practice, and considers the factors affecting designing and producing in design projects	Chapters 1, 2 and 3
P2.1	Identifies design and production processes in domestic, community, industrial and commercial settings	Chapter 4
P2.2	Explains the impact of a range of design and technology activities on the individual, society and the environment through the development of projects	Chapter 5
P3.1	Investigates and experiments with techniques in creative and collaborative approaches in designing and producing	Chapter 6
P4.1	Uses design processes in the development and production of design solutions to meet identified needs and opportunities	Chapter 7
P4.2	Uses resources effectively and safely in the development and production of design solutions	Chapter 8
P4.3	Evaluates the processes and outcomes of designing and producing	Chapter 9
P5.1	Uses a variety of management techniques and tools to develop design projects	Chapter 4
P5.2	Communicates ideas and solutions using a range of techniques	Chapter 10
P5.3	Uses a variety of research methods to inform the development and modification of design ideas	Chapter 7
P6.1	Investigates a range of manufacturing and production processes and relates these to aspects of design projects	Chapter 4
P6.2	Evaluates and uses computer-based technologies in designing and producing	Chapter 10
H1.1	Critically analyses the factors affecting design and the development and success of design projects	Chapter 11
H1.2	Relates the practices and processes of designers and producers to the major design project	Chapter 12
H2.1	Explains the influence of trends in society on design and production	Chapter 13
H2.2	Evaluates the impact of design and innovation on society and the environment	Chapter 14
H3.1	Analyses the factors that influence innovation and the success of innovation	Chapter 15
H3.2	Uses creative and innovative approaches in designing and producing	Chapter 16
H4.1	Identifies a need or opportunity and researches and explores ideas for design development and production of the major design project	Chapter 20
H4.2	Selects and uses resources responsibly and safely to realise a quality major design project	Chapter 20
H4.3	Evaluates the processes undertaken and the impacts of the major design project	Chapter 20

Syllabus Outcomes	Study guide chapters
H5.1 Manages the development of a quality major design project H5.2 Selects and uses appropriate research methods and communication techniques	Chapter 20 Chapter 20
H6.1 Justifies technological activities undertaken in the major design project through the study of industrial and commercial practices H6.2 Critically assesses the emergence and impact of new technologies, and the factors affecting their development	Chapter 20 Chapters 17, 18 and 19

Each of these chapters is filled with course content and activities to develop your level of experience and extend your knowledge. HSC style questions and answers provide opportunities for practise and improvement.

Planning your own learning

This study guide can be used in several ways:

- As a self-contained section at the beginning of a new topic to prepare skills before they are covered in class.
- To build skills when you are working on assignments throughout the course.
- As a reference to guide you through the course when the need arises.

The summary cards included in the removable section at the back of the book provide a short summary of the information in the guide. They are intended to be used for revision before tests and exams as they provide an overview of the course.

Looking ahead

To maximise the features of this study guide you need to set clear goals. Try to consider the following when using the guide:

- What are my strengths and weaknesses?
- Which practical skills do I need to practise for my major design project and how will I use my time management skills effectively?
- How will I find and apply the technology used in my project?
- How will I build on my creative thinking and problem-solving skills to improve my design skills?
- How will I document my findings and communicate them to a wider audience?

Setting goals

A key to success is identifying and setting goals and then checking them later to see that they have been achieved. This helps you to plan and monitor your progress. You will need to develop an individual action plan for each chapter in the study guide. To do this, consider the following questions:

- Where am I now? Here you will need to evaluate the skills achieved by identifying your strengths and weaknesses, investigate and utilise a range of research techniques, collect data and justify the level and standard at which you are working.
- Where do I want to be? Once a skills audit is completed you must set some SMART goals to determine the level which you desire to achieve and the steps that you will follow to get there.

The acronym SMART is often used to set goals. It is used to check that goals are Specific, Measurable, Achievable, Realistic and Time constrained.

Specific	Clearly identified a small number of related tasks
Measurable	Consider how you will know that you have achieved the goal
Achievable	It must be possible to achieve the goal considering resources, time, cost and skills and abilities
Realistic	Not too optimistic or challenging given the time constraints
Time constraints	When will the goal be achieved? How long will it take? When can you start?

What are my next steps?

Achieving goals requires planning. When following a plan you can always see how much you have progressed towards your goal and how much more input is required. Knowing where you are is essential for making good decisions on where to go or what to do next.

How successful was I?

Imagine how much easier your life would be if you were able to identify the specific and measurable actions you needed to make your work a success. To evaluate your own work, develop written evaluations as key points throughout your project. Evaluation should be a continuous ongoing process.

You will need to regularly evaluate the planning process, the management and use of time, the research methods employed and the difficulties encountered. The quality of information found will directly relate to the research methodology used and the method of recording research data. The creativity and innovation in your work and its progression and design development will ultimately be determined by the quality of the design decisions made and the effective choice of manufacturing technologies.

Study skills

How do people learn best? We learn through a range of techniques and some of us are predominantly visual, auditory or kinaesthic learners. Each of us must determine which learning techniques we benefit from the most and use them to our advantage. The following table demonstrates the percentage of the population that benefits from the stated techniques.

5%:	Lectures
10%:	Books
20%:	Audio-visual aids
30%:	Dramatic lecture with audio-visual aids
50%:	Discussion of idea with other people
75%:	Explaining it to someone else
90%:	Teaching it to someone else (Ekwall and Shanker, 1988)
95%:	Assessing someone else (Glasser, 1990)

The research data from the above table tells us that the majority of the Australian population learn best when they teach the topic to someone else and assess their learning. Within the classroom, you could work with a peer who is having trouble understanding a concept to teach them what you know.

Using what you know

It is essential to develop and build upon a set of key skills. These are the skills that are necessary to improve the quality of your work. Key skills may include:

- good communication and literacy skills
- numeracy
- Information Communication Technology (ICT) skills
- working effectively in teams
- problem-solving and decision-making skills
- flexibility and adaptability to change
- time and financial management approaches
- creativity
- technological skills
- knowing which learning methods work best for you
- ongoing evaluation.

The Australian Design Council believes that 'design thinking is transferable thinking' because the attributes we use to learn new things are the same as design learning attributes.

Ensuring quality work

Continuous improvement is the fundamental principle here. This means that students should be constantly practising their skills, evaluating their work and always working towards improving the quality of their output. Quality work involves reaching the highest possible standard achievable. 'Quality' can be defined as that which satisfies and exceeds customer needs and wants. When working on the major design project it is necessary to set a series of checkpoints where quality and standards can be checked against relevant criteria at appropriate times.

Research

A vital component of your success in this course is research. Your research and data analysis will underpin your entire design project. The information-gathering process will require careful planning, thorough analysis, documentation and application. It is important when gathering data to use both primary and secondary sources and, when analysing the data, to use both qualitative and quantitative analysis methods.

Primary research involves collecting your own data from a range of sources including interviewing experts face-to-face or by telephone; communicating with clients; fieldwork (physically collecting data); producing questionnaires and surveys; exhibitions and displays; and testing, experimenting and building prototypes, models and computer simulations. Primary sources are your original contributions. The best research is based on information that you have collected first-hand. Judgements will have to be made and recorded based on this data.

Secondary research sources are easier to obtain than primary sources and can provide solid background information. Sources of secondary research include articles from books and journals; items from catalogues, magazines and multimedia; and printed matter such as flyers and brochures, reports and Internet material. The latter must be checked for validity and should be sourced through reliable search engines such as Google Scholar.

Many students include irrelevant research data in their folios. To ensure you have the best opportunity to gain the highest marks only include information that impacts directly on any decision making in the development of your design solution. Include those research findings that will affect project decisions.

In order to proceed:

- identify the information needed
- plan your information gathering
- use a range of primary and secondary sources
- consult experts
- keep accurate records
- evaluate research data and justify your reasons for decisions made about the MDP
- include tests, modelling and experimentation
- report the results using qualitative and quantitative research techniques.

Qualitative research refers to in-depth responses from individuals, such as statements based on opinions. Quantitative analysis refers to statistical analysis using numbers, percentages and scores or scales that are used to transfer values into numerical data.

Partnerships

The development of partnerships with mentors can be very beneficial in the Design and Technology course. A mentor is a person who has expertise in an area that you are interested in and is willing to share that expertise and experience with you outside of class time in support of your major design project.

You must get permission from your parents or guardians and your teacher to work with a mentor. Your teacher may be able to suggest a suitable mentor, but as a matter of routine if you are working with mentors, your teacher must organise a child protection check to be completed by your local police before you start working with them. To do this, your teacher will provide you with the child protection form, supplied by your school. Your mentor must complete and sign the document. It must then be returned to the school who will forward it to the police for processing. If there is an issue or the mentor has a record, the school will be contacted. This is a legal requirement in NSW and incurs no cost to you or the mentor.

Preliminary course

1–Design theory and practice

PRELIMINARY OUTCOMES

A student:	You learn about:	You learn to:
P1.1 examines design theory and practice, and considers the factors affecting designing and producing in design projects.	■ design theory and practice – the range of design professions – the nature and variety of work of a range of design professions – the interaction and overlap of design professions – Australian and international designers and their work.	■ investigate at least one designer and the nature of their work. ■ identify a range of career opportunities in design and production.
	■ design processes – the design processes used in domestic, community, industrial and commercial settings from initial contact with clients to final presentation.	■ describe and analyse the processes undertaken when designing. ■ apply a design process when developing design projects.

1.1 Design theory and practice

The importance of integrating higher-order thinking, creativity, innovation and design as a conceptual tool to solve problems in technology education cannot be underestimated. This is the premise behind design theory and practice. If design is not taught and remains a 'poor relative' of technology then design thinking will not develop.

The challenge for technology students is to move their thinking from that of a student involved in traditional skill-based procedural knowledge to one who uses and promotes design as a tool to solve problems. Design is not solely about aesthetics or appearance: it is a concept, a problem-solving process that can be applied to almost all situations. The importance of design and the value of creative thinking when solving problems cannot be underestimated. Designing is a complex intellectual activity that requires higher-order thinking skills: it requires a conceptual change.

The terms 'design' and 'technology' appear to contradict each other. Design refers to a creative synthesis of ideas using a design process; technology is the repetitive use of tools to manufacture an item. Design is a conceptual tool used to get people to think about why we make things the way we do, to evaluate the best construction methods, and to synthesise new ways of doing things. Design processes are used to create synthesised, original and purposeful solutions to problems. Design is a tool that allows individuals to make meaning from what is around them and its use promotes lifelong learning. Historically, technology was driven by manual skills and followed a skills-based vocational approach that focused on the ability to mimic or copy a set skill.

Design is central to Australia's international competitiveness. It is essential to the process of innovation and therefore central to the commercial success of innovation investments. It is critical that design must operate at all levels of production, from idea conception to realisation. An increased community awareness of the role and benefits of good design will create more discerning consumers. Design is integral to the provision of goods and services and can uniquely shape the environment in which we live. Because of this there is an increasing call for all design to have an environmental conscience that lowers our generational ecological footprint.

Design should be integrated through brainstorming initial ideas, researching the problem in-depth using primary and secondary research, producing final sketches, choosing and using manufacturing materials, tools and techniques, and ongoing evaluation.

Technology education, on one hand, requires low-level intellectual skills and is used as a vehicle for developing physical skills in manipulating concrete materials but on the other hand, because the activities are challenging, more profound learning can occur (Middleton, 2005). When creating solutions to real problems designers must use early exploration and investigation of existing ideas, tools, techniques and materials. This allows designers to gain a complete picture of the problem, the existing solutions and emerging technologies and then brainstorm the entire range of possible solutions, each of which can be developed and improved upon until a final direction is obtained and a solution found.

Who is a designer?

A designer is a business professional who develops solutions to commercial needs that require the balancing of technical, commercial, human and aesthetic requirments.

Designers must balance the needs of their clients with the needs of the intended users of the design. These are often the employer's customers. Designers must reconcile their own standards of aesthetics, quality and ethics with the requirements of the intended commercial purpose of their work. Both designer and client should also consider community, cultural, environmental and ethical values and constraints.

Designers must be able to cope in a world where change is constant. The concept of design will lead us into the next decade and in turn teach designers to welcome change, be creative and innovative and solve problems as they arise. Most importantly, designers must be autonomous problem-solvers who will be an asset to our country. Australia has an important role to play in global design imperatives.

1.1.1 The range of design professions

While there are a range of design professions, the role of the designer across these professions is a common one: using the design process to solve problems.

1.1.2 The nature and variety of work of a range of design professions

Industrial design

Industrial designers develop and prepare products for manufacture. They relate human usage and behaviour to product appeal.

Industrial designers are also known as product designers. They explore solutions to meet marketing, manufacturing and financial requirements. They consider both functional and aesthetic aspects and pay attention to ergonomics: those factors that relate to ease of use and human behaviour.

They prepare models and prototypes as well as drawings and illustrations to assist in the decision-making process and support marketing efforts. They select components and materials, resolve assembly and manufacturing details and produce digital and documentary instructions for others involved in the manufacturing process.

Industrial designers may work in manufacturing companies, design consultancies, as designers/makers in workshops, in research and development, or be self-employed.

Interior architecture and design

Interior designers plan and detail commercial and residential building interiors for effective use, with particular emphasis on space creation, space planning, and factors that affect our responses to living and working environments.

Interior design is also referred to as interior architecture because interior designers are trained to consider the modification of the interior structure of the building rather than just refinishing and furnishing existing spaces.

Interior designers plan space allocation, traffic flow, building services, furniture, fixtures and surface finishes. They consider the purpose, efficiency, comfort, safety and aesthetics of interior spaces to arrive at an optimum design. They custom design or specify furniture, lighting, walls, partitions, flooring, colours, fabrics and graphics to produce an environment tailored to a purpose.

An interior designer often works as part of a team that may include architects, builders, project managers, engineering consultants, shopfitters, cabinet makers, and furniture and materials suppliers. They may be required to organise the modification of building structures, the purchasing of materials and furnishings and the contracting and supervision of the tradespeople required to implement a project.

Interior designers work in interior design consultancies, architects' offices, as builders and in government building and construction departments. They may be employed as shopfitters or by shopping centre corporations, be self-employed, work for building industry suppliers as furniture designers, be exhibition designers, or work on film, television and stage set design.

Graphic designers

Graphic designers develop and prepare information for publication. Graphic designers refer to their area of specialisation as visual communication.

This profession requires a sound understanding of text-based communication, including the use of symbols, colours and images. They prepare concept layouts and mock-ups for clients and produce or subcontract diagrams, illustrations and photography. They resolve all communication elements into a final format to suit the required physical or digital media.

They select paper and other printing materials, resolve manufacturing details and produce instructions for others involved in the reproduction process. They organise and oversee proofs and colour separations to prepare for printing, and liaise with suppliers.

The volume of visual material produced to support both commercial and cultural purposes means that the areas of employment open to graphic designers are broad. The rapidly developing areas of digital media relating to the Internet and multimedia business presentations are opening up new areas of employment for graphic designers.

Work for graphic designers can be found in advertising agencies, public relations firms, design consultancies, government, business and manufacturer's promotional departments, desk-top publishing, web design, printing or packaging companies, book publishers, newspapers, magazines or in self-employment.

Textile designers

Textile designers plan and develop textiles, fabrics and patterns. They are also involved in the manufacturing techniques of fabrics and other patterned materials. They design the structure of the fabric and make decisions about appropriate yarns, colour use, surface patterning, textures and finishing.

Textile designers develop materials used in furniture, clothing, vehicles and products such as luggage. They can apply the same skills to the development of patterns for wallpapers, laminates and patterned plastics.

They design fabrics to satisfy marketing and manufacturing requirements, balancing aesthetic and functional aspects and considering the nature of yarn types, thicknesses, weights and textures to produce fabrics to cost and production constraints.

They prepare design concepts and assess them for market viability. They advise and liaise with industries where it is necessary to predict future colour trends. They monitor trends in industries such as interior design, automotive design and fashion, and progressively evolve fabric styles to meet these specific needs.

Textile designers work in clothing companies, textile manufacturing companies, design consultancies, designer/maker workshops, for fabric wholesalers, in automotive colour and trim departments or are self-employed.

Exhibition and display design

Exhibition designers design and organise the construction and installation of trade exhibitions, permanent shop displays, museum exhibits and interpretive displays. They use skills drawn from graphic, industrial and interior design to attract, inform and involve an audience in the subjects that their clients employ them to present.

These designers work in exhibition and trade display businesses, in museum and cultural institution design teams, in interior design consultancies (specialising in museum work or retail shop interiors), department store display departments, in theatre, film or television set production, or are self-employed.

Fashion design

Fashion designers develop clothing, accessories, footwear and other items of personal apparel. They study the design and construction of clothing, its historical development and styles, and the techniques and processes available for its manufacture. They rely heavily on illustration skills and the making of samples to communicate their designs.

They prepare designs to meet marketing, manufacturing and financial requirements. They consider both functional and aesthetic aspects and pay particular attention to the relationship of the apparel to the human form.

They prepare clothing samples to demonstrate and test products, and drawings and illustrations to assist in the decision-making process and to support marketing efforts.

They select materials, resolve assembly and manufacturing details and produce patterns for others involved in the manufacturing process. They prepare the product for production and develop and oversee subsequent adjustments and refinements.

It is common for fashion designers to be employed within manufacturing companies who develop mass-produced clothing for the retail trade. Others work in small businesses, making specialised lines of clothing for boutique retailers or producing custom-tailored clothing for personal clients.

Television, film and theatre set design

Set designers plan and manage the construction of sets for the presentation of theatre, television and film productions. Set designers must understand the production requirements of the entertainment media they are designing for and pay attention to methods of assembly and disassembly, as well as strength and safety aspects.

Set designers plan and detail performance stages and sets for effective use, with particular emphasis on space planning, factors that affect the production of the performance, and visual impact.

A set designer often works as part of a team that may include lighting designers, sound technicians, choreographers, directors and cinematographers. They may be required to organise the purchasing of materials and furnishings and the contracting and supervising of any tradespeople involved in a project.

Work can be found in theatre and film companies, television studios, advertising production companies or in corporate events production.

Design education

Design education has become a major growth area in both secondary and tertiary education. The education of designers requires teachers with knowledge in the many subject areas that designers must study. It also requires experienced designers in each design discipline who are able to pass on their knowledge of professional practice.

Design educators may have qualifications in a design discipline or in one of the subjects that make up the curriculum. They may additionally have teaching qualifications.

Jewellery design

Jewellery designers conceptualise, prototype and detail items for manufacture. They have specialised knowledge of the metals, precious stones and other materials associated with personal adornment. They may develop designs for mass or batch production or they may develop special items to satisfy one-off commissions. They may also design other objects that use precious metals and jewelled decoration such as trophies and cutlery.

They prepare models and prototypes as well as drawings and illustrations to assist in the decision-making process and support marketing efforts. They select components and materials, resolve assembly and manufacturing details and, if they are working in a mass-production environment, provide instructions for others involved in the manufacturing process. They also oversee subsequent adjustments and refinements to the product.

Jewellery designers may be employed by manufacturing companies specialising in jewellery and other decorative wares. They may also work in jewellery shops where small batch production or repairs and adjustments are done. Some operate as independent designer/makers, producing custom designs or batches of jewellery for small retail outlets.

Furniture design

Furniture designers develop and prepare furniture for manufacture. Furniture design is a specialist area of industrial design. Furniture design is also undertaken by interior designers and, traditionally, by architects. These designers may be seeking a unique style of furnishing for an interior or architectural project.

Furniture design has a rich history and a close relationship with fashion, which makes this area distinctly different from many product design fields.

Furniture designers explore solutions to meet marketing, manufacturing and financial requirements and arrive at the optimum design of a furniture item. They consider both functional and aesthetic aspects and pay particular attention to ergonomics. They prepare models and prototypes as well as drawings and illustrations to assist in the decision-making process and support marketing efforts. They select components and materials, resolve assembly and manufacturing details and produce digital and documentary instructions for others involved in the manufacturing process. They organise tooling to prepare for production and develop and oversee subsequent adjustments and refinements to the furniture.

Employment will be found in manufacturing companies, design consultancies, designer/maker workshops or as self-employed designers.

Multimedia and web design

Multimedia and web designers develop and prepare information for digital publication. This requires a sound understanding of text-based communication and requires them to skilfully use the communication properties of symbols, colours, pictures, animation, video and sound.

They prepare concept layouts and mock-ups to discuss project details with clients. They prepare or subcontract diagrams, illustrations and photography and resolve all communication elements into a final format to suit the required digital media.

Digital designers often work as part of a project development team. While specialists such as copywriters, photographers, illustrators, sound recordists, composers and animators are preparing specific elements of the production, the multimedia or web designer may be working on the overall presentation. This involves the layout of each page or section and the navigation methods that allow the user to access the publication.

Work can be found in advertising agencies, public relations firms, design consultancies, government, business and manufacturer's promotional departments, web design, digital production companies, manufacturer's promotional departments, publishers, newspapers, magazines and via self-employment.

1.1.3 The interaction and overlap of design professions

Designers may choose to work as freelancers managing their own businesses or work for a design company as a part of a design team.

Freelancers

Freelance designers are often able to choose their clients and projects as well as set their working hours. They must have an excellent reputation because their business generally relies on word of mouth recommendations and Internet marketing.

Design teams

Design teams exist within both small and large companies. It is common for larger companies to have a number of design teams with specialised expertise to cater for each client's specific needs. The process they usually follow is detailed below. Note that these processes may occur concurrently.

1. Initially the client will meet with the chief designer. The client will provide a description of the need, along with background information and all limitations. Discussions will include an analysis of preferred functional and aesthetic properties as well as preferred materials, tools, techniques and design styles.
2. The chief designer will summarise this data into a design brief and the client will confirm that the brief is clear. The chief designer then presents the brief to the design team, usually consisting of three to four individuals, who will brainstorm or brain-jam as many solutions as possible.
3. These solutions are divided up between the team who will investigate which are viable. The team then meets and presents their findings. Collaboratively, they decide which solution/s will undergo further research. The person who put forward the solution will then lead further investigation, dividing tasks between the team with each member reporting back to them.
4. The design process initially involves drawing and improving the idea using thumbnail sketches. Each idea needs to be researched in-depth, using both secondary and primary research.
5. Secondary research will include an investigation of the problem: a critical evaluation of existing solutions is the first step. This investigation is extended by examining the impact of changing the possible design styles as well as the use of materials, tools, techniques and emerging technologies.
6. Primary research will be carried out. This will involve interviews, surveys and questionnaires of the target market, and testing and experimentation with design ideas, materials, tools and techniques.
7. Qualitative and quantitative data will be thoroughly evaluated and analysed before a final sketch is produced. Qualitative data deals with subjective feelings about likes and dislikes and quantitative data provides information on objectively measured statistical data.
8. A final production sketch will be produced showing every aspect of the final design solution. A high-quality prototype is produced to market the idea to the client.
9. The chief designer will request the design team to present both the folio and the prototype for a final evaluation. Communication techniques are of the utmost importance at this stage.
10. The client will make the final decision as to whether the solution will go to manufacture, usually by being outsourced. Successful designers possess some knowledge of technology, manufacturing, science and engineering as the greater understanding the designer has of the materials, tools and techniques to be used in the manufacturing process, the more successful the design solution will be. Designers also have some understanding of sociology and psychiatry and often develop strong bonds with clients.

1.1.4 Australian and international designers and their work

With the ever-improving global communication systems through satellite and Internet technology the world has become a much smaller place, resulting in designers from all over the world collaborating to create products, systems and environments. The wings of Boeing aircraft are designed in the western suburbs of Sydney and the plans and models transmitted to a 3-D photocopy machine in Seattle where they are manufactured and attached to the plane.

Designers, like computers, are not stand-alone versions. They are networked all over the world so Australian designers no longer live in isolation and are able to access the latest and best materials, tools and techniques available, which in turn helps them to create best design practice.

Case study 1: Graphic designer Kate Jones

Job description

Kate Jones is a graphic designer and mainly works on print-based designs such as advertisements and brochures. She also works on concepts for new brands and products, which can include developing logos, direct mail concepts or exhibition stands: basically, whatever is asked of her. She is currently employed by The Never Board Room at Clock Creative Communications in England, where she moved from Australia to further her career opportunities and experience.

Educational qualifications

Kate completed a foundation course and a BA (Hons) in Graphic Design. On completion she did not feel she was ready to enter the workforce so she went to University College, Falmouth and completed a post-graduate Diploma in Creative Advertising. This additional year provided Kate with the experience of what it would be like to work in a real studio as the class presented work to each other and had agency-style deadlines.

Kate is known as a people person and this gives her the confidence to go into meetings and present ideas. She is organised, which is essential when you have to deal with competing deadlines in addition to daily phone calls and meetings.

Inspiration

Kate is inspired by other designers: both the designers who are designing in the same room as her and others beyond. She loves the crazy 'out there' ideas she sees in magazines and on the web and understands that the simplest ideas can be the hardest to come up with.

'A typical day'

There is no typical working day as such. In her current role as the lead designer and account handler at Clock Creative Communications she is responsible for several clients. This means that any of them might call with a job that needs doing or a question that requires some research. She might put together quotes, work on ideas or projects and then attend a client meeting to discuss forthcoming work.

At Clock they regularly have brainstorming sessions: 'they can be for any of our clients and we all attend, from the office manager to the MD'. They also run evening sessions called Thinking Cap where the client is invited to discuss where their brand is going or generate ideas for a specific project. This is a great opportunity to work creatively outside of the work environment.

The company

In the studio there are four designers and the creative director. Kate enjoys the fact that every day is different and that she is involved in projects from their beginning to completion.

Advice to others

Kate advises aspiring designers to gain as much work experience as they can. It is impossible to get real-life experience at university, especially with aspects such as dealing with printers and discussing costings with clients. Kate also recommends that students make sure they know their software, as employees expect a working knowledge of design programs that are used on a daily basis. New designers need to have some professional pieces in their portfolio, and be confident when they talk about their work. Networking can also be useful for opening doors and getting known in the industry.

The future for Kate and graphic design

In the future Kate can see herself doing much of the same thing but would love to work on national campaigns. The main issues currently affecting the graphic design industry are the increase in digital work over the more traditional print and television areas, and whether there is room for both disciplines. It is important to learn web skills to maximise all the opportunities that come along.

Case study 2: Fashion designer Peter Alexander

Nature of work

Peter Alexander is an Australian fashion designer specialising in sleepwear and accessories. His name has become synonymous with the quality fashions he creates but he has said that he created a huge business out of a 'stupid idea'! His company, Peter Alexander, is currently owned by the Just group but he continues in the role of manager.

Educational history

At high school Peter was voted the least likely to succeed and one of his teachers told him that he should leave school and go to TAFE. Not liking being told what to do he stayed at school and successfully completed Year 12. He states that doing this gave him the confidence that he needed to succeed. Peter's only regret about his education is that he did not do a business course. He says that 'as a result I didn't make any money from my business for seven years because I didn't know what I was doing'.

Training and inspiration

Peter has no formal training and no business experience, both of which have been stumbling blocks that he has worked hard to overcome. Because of this lack of experience many people discouraged his endeavours.

With no formal design training Peter does little sketching, instead making lists of ideas that catch his imagination. Words such as 'art deco', 'diamantes', and so on are written on the page and then Peter uses an illustrator to sketch his ideas prior to manufacture.

Professional history

Peter started his career in retail, working there until he was twenty-four years old. After listening to his female friends he identified a gap in the fashion market for comfortable pyjamas that were not made of flannelette or looked like 'granny' pyjamas.

When a department store called and cancelled an order for 2000 items Peter was devastated but after some quick thinking his biggest setback became his greatest opportunity: he immediately placed a mail-order advertisement in *Cleo* magazine and had 6000 orders within three weeks. After this there was no looking back.

A few years ago, when he thought he had taken the business as far as he could, Peter sold the company to the Just group. At the time of the sale the business was making six million dollars per year and growing at an annual rate of forty per cent. Peter will continue to run the business as part of the sale arrangement.

Believing in the importance of setting high goals he says, 'people with high goals have a much better chance of achieving them'. He also says that the most satisfying thing about having your own business is being able to take both the blame and credit for your own work.

Peter currently manages the company, which employs eight full-time and twenty casual staff. In 2008 Just took the company global, opening over thirty stores in New Zealand, New York and Los Angeles.

Products produced

Initially the only products produced were pyjamas: summer and winter, male and female. The range has now increased to include anything from embroidered sheepskin boots to an entire summer collection that merges sleepwear with casual clothing items.

Nature of the designer's work

When Peter initially started the business he worked closely with his mother to get the enterprise off the ground. They did everything themselves including designing the clothes; sourcing and selecting fabrics; manufacturing the patterns; grading them to the different sizes; manufacturing the items and their labels and packaging; designing and distributing postal order catalogues to retailers; and recording and distributing completed orders. As the business has expanded Peter has hired full- and part-time staff to take on many of the tasks.

Factors contributing to the designer's success

Peter will tell you that a strong determination, often described as stubbornness, is his greatest strength. He has very supportive parents who believe that he is capable of anything and they have provided ongoing support throughout his entire career.

Factors limiting the designer's success

Peter claims that his greatest limitation was his lack of knowledge and training in business. He believes he could have been financially successful much sooner if he had business training. Formalised training as a designer would also have been advantageous, especially in the sketching of designs.

1.2 Design processes

Design is a process used to solve problems. The type of design process used will depend on the degree of difficulty in the problem or design brief. A design process can be simple and the problem-solving process automatic. It is a process of decide, evaluate and act. For example, when deciding what time to get out of bed each day a person will consider their diary and the tasks to be completed, and set their alarm accordingly.

Design processes require some research into the problem. Sketching initial ideas; researching the problem; testing and experimenting with materials, tools and techniques; sketching a range of evolved, informed ideas; evaluating these ideas and selecting the best; and manufacturing and evaluating the final solution are all part of the process. For example, when deciding what clothes to wear each day an individual will examine the resources available (what is clean, what is weather appropriate, what is activity appropriate) and make an informed decision.

Design processes can be formalised between a client and a designer or design team. The client will provide the designer with a problem, their initial ideas and expectations, and a list of limitations: time, cost, energy and materials. This information must be discussed, clarified and formalised in a documented design brief.

This will involve ongoing meetings where communication must be clear and firm decisions made until the solution is realised. Initially the designer will present a range of possible solutions to the client to determine their preferences. The importance of concept sketches and clear communication cannot be underestimated.

Once a clear direction is found the designer will research existing solutions and new and emerging materials and technologies that may be used. This will impact on the evolving design solutions which are then presented to the client, allowing them to narrow the direction that the designer will move in. The designer will then test and experiment with materials and manufacturing processes, making samples and prototypes to discuss with the client.

Once the design, materials and manufacturing preferences are selected the designer will again examine the budget to ensure the design is feasible and discuss this with the client, who will approve all or part of the design. If the client rejects any part of the design the designer will need to evaluate the progress to date and research further options. Once the client and designer are in agreement the solution is manufactured and its success evaluated.

As this process may occur over a long period of time, and the designer can be working with a number of clients at once, it is important to keep clear records. These records are maintained in the form of a design folio specifically following each of the steps in the design process.

1.2.1 Design processes used in domestic, community, industrial and commercial settings from initial contact with clients to final presentation

In the domestic setting the design process is not formalised. Within the community the design process used will depend upon the people involved and the type of decisions being made. Some will be formal and others informal.

In industrial and commercial settings the design process is always documented in a design folio and is a formal process. This aspect is fully detailed in Chapters 7 and 16 of this study guide.

Key concepts and definitions

Design is about problem-solving, innovation and emerging technologies. Design is not solely about aesthetics or appearance: it is a concept, a problem-solving process that can be applied to almost all situations.

Designer—a business professional who develops solutions to commercial needs that require the balancing of technical, commercial, human and aesthetic requirements. Designers must be able to cope in a world where change is constant.

Design process—used to create synthesised, original and purposeful solutions to problems. Design is a tool that allows individuals to make meaning from what is around them. The use of design as a tool promotes lifelong learning.

Design professions—there are a range of design professions and the role of the designer across these professions is a common one: designers are problem-solvers and design is about using the design process to solve problems.

Technology—the use of tools to complete the manufacture of an item.

Useful websites

- architecture.about.com
- eng.archinform.net
- pod@marc-newson.com
- www.architecture.com.au
- www.collettedinnigan.com.au
- www.gagosian.com
- www.glassbrasserie.com.au
- www.greatbuildings.com/architects/Glen_Murcutt.html
- www.henrijosef.com.au
- www.lukemangan.com.au
- www.lukemangan.com.au/pdf/lukes_bio.pdf
- www.marc-newson.com
- www.PeterAlexander.com.au
- www.tigerlilyswimwear.com.au

Classroom activity

1. Use the following table to gather data to allow you to complete a comparative case study on an Australian and an international designer.

Designer	Company name	Contact details	Nature of work	Possible resources
Collette Dinnigan	Collette Dinnigan	Collette Dinnigan Pty Ltd 22-24 Hutchinson Street Surry Hills NSW, 2210 Ph: 02 93610110 Email: mail@collettedinnegan.com.au	Fashion and accessories	www.collettedinnigan.com.au
Jodhi Meares	Tigerlily	Head Office Level 1, 267 Cleveland St Surry Hills NSW 2010 Ph: 02 82288000	Swimwear	www.tigerlilyswimwear.com.au
Luke Mangan	Glass Brasserie, Sydney Hilton Hotel	Sydney Hilton Level 2, 488 George St Sydney Ph: 02 92656069 Email: eat@glassbrassserie.com.au	Food and menu designer	www.lukemangan.com.au www.lukemangan.com.au/pdf/lukes_bio.pdf www.glassbrasserie.com.au
Henry Roth	Henry Roth, Project Runway	Ph: 02 92111733 Email: henry@henrijoseph.com.au	Wedding gowns	www.henrijosef.com.au
Peter Alexander	Peter Alexander	Ph: 1300366683 658 Church St Richmond Victoria	Pyjamas and accessories	www.PeterAlexander.com.au
Marc Newson	Marc Newson	Marc Newson 7 Howick Pl London SW1 1BB UK Skype: +442079320990	Industrial and furniture designer	www.marc-newson.com www.gagosian.com pod@marc-newson.com
Glenn Murcutt	Glen Murcutt	Glenn Murcutt Level 2 7 National Circuit Barton ACT Ph: 02 61212000	Architect and environmentalist	architecture.about.com eng.archinform.net www.greatbuildings.com/architects/Glen_Murcutt.html www.architecture.com.au
George Gross	George Gross George Gross and Harry Who	Sydney Office 45 Cross Street Double Bay NSW 2028 Ph: 02 93262088	Fashion	www.georgegross.com.au
Georg Jenson	Georg Jensen We	George Jensen Pty Ltd 18/3 Vuko Place Warriewood NSW 2102 Ph: 02 99134203	Jewellery, flatware and silverware	www.georgjensenstore.com.au www.jensensilver.com

Designer	Company name	Contact details	Nature of work	Possible resources
Catherine Martin	Catherine Martin	Catherine Martin Pty Ltd PO Box 430 Kings Cross NSW 1340 Ph: 02 9334555 Email: reception@catherinemartin.com	Costume designer and television producer	www.catherinemartin.com
Marcel Wanders	B&B Italia, Bisazza, Poliform, Moroso, Flos, Boffi, Cappellini, Droog Design and Moooi	Marcel Wanders Studio Westerstraat 187 1015 MA Amsterdam, The Netherlands Ph: +31 20 4221339 Email: joy@marcelwanders.com	Industrial designer, architectural, interior and home appliance designer	www.marcelwanders.com
Jasper Morrison	Jasper Morrison Alessi	www.jaspermorrison.com	Industrial, furniture and product designer	www.jaspermorrison.com www.designboom.com
Maurice and Paul Marciano	Guess by Marciano	www.guess.com	Fashion and accessories	www.marciano.com www.guess.com
Tamara Melion	Jimmy Choo	Skype: +442073685000 Email: customercare@ jimmychooonline.com	Shoes, handbags, sunglasses, accessories	www.jimmychoo.com

Information on designers can be gathered from the resources listed by completing an Internet search on the designer and/or their company, writing to the designer/company address, telephoning the company receptionist, using your local, state or university library and examining design journals, books, texts, DVDs and multimedia materials.

Key concept question A p. 196

Choose one of the designers above. Download, read and highlight or summarise the information on the designer into a case study by doing the following:

1. Write down the name of the selected designer and the company that they are associated with.

   List the range of products that this company produces.

   Compile a resume on your designer. Include information on their background, education and training, career history, their design inspirations, and factors contributing to and/or limiting their success.

   List the nature of the work that your designer currently does.

   Describe what this work involves and why it is important.

Sample Preliminary questions

Extended-response questions A p. 196

1. Justify the role of design within a problem-solving, experiential design process. (15 marks)
2. Design and technology are a 'contradiction in terms'. Explain how the two come together and support each other in the design process. (15 marks)

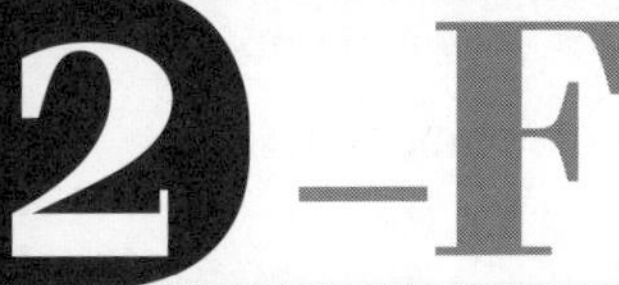

2 – Factors affecting design

PRELIMINARY OUTCOMES

A student:	You learn about:	You learn to:
P1.1 examines design theory and practice, and considers the factors affecting designing and producing in design projects.	■ the factors affecting designing and producing including – appropriateness of the design solution – needs – function – aesthetics – finance – ergonomics – occupational health and safety – quality – short- and long-term environmental consequences – obsolescence – life cycle analysis.	■ identify factors affecting design – analyse design products – compare and contrast the factors to be considered in the design and production of design projects – appraise the aesthetic and functional qualities of a variety of design products, systems and/or environments.

2.1 The factors affecting designing and producing

The factors that impact on designing and producing include the appropriateness of the design solution; the need that is being fulfilled; the function of the product, system or environment; the aesthetics of the design; the cost; ergonomics; occupational health and safety; quality; environmental consequences; obsolescence; and life cycle analysis.

2.1.1 Appropriateness of the design solution

A designer must determine the appropriateness of a design solution. To do this they must consider whether the solution meets the need and the criteria for success. To initially set the criteria for success, the designer must list all important factors and then prioritise and rank these.

Design questions to be considered may include the following:

- Does it solve the problem?
- Is the solution functional, efficient and easy to use?
- Is it aesthetically pleasing?
- Is the solution within the client's limitations of cost, price and time?
- Does it have a minimal ecological footprint?
- Does it use sustainable materials?

2.1.2 Needs

In society a 'need' may be defined as something one must have in order to survive and a 'want' is something one would like but is not essential for survival. In the design world a need is seen as a problem that requires solving.

An example of a design problem may be a showerhead which needs replacing. The client wants one that provides a strong flow but also saves water. The new showerheads on the market save water but do not provide a strong enough flow. The design need then becomes to create a water-saving showerhead that produces a strong flow of water.

Perceived needs are things that are not essential to sustain life but people feel they cannot live without. This is how new designs are marketed, with the advertising agency convincing the consumer they must have this new creation. Individuals believe they need the product in order to:

- belong to a group. For example, if they use a piece of sport equipment they will be part of the group of elite athletes who also use the product.
- identify with a role model. For example, when a celebrity endorses a product the consumer feels they can identify with their achievements.
- be healthy by consuming or using certain products.
- maintain their status by using certain brands associated with socioeconomic status.
- keep up with their peers by purchasing the latest gadgets and fads.

2.1.3 Function

The function of a design refers to how and whether it works for its intended use. Function may be assessed by examining the safety factors; the strength of the product; the ease of use and clarity of instructions; the degree of efficiency; the durability of the materials used and their short- and long-term environmental impact; their obsolescence; and a life cycle analysis.

2.1.4 Aesthetics

The aesthetics of design is very important. Aesthetics refers to the overall appeal of the creation. Many students confuse aesthetics with design: design refers to the process used to solve the problem; aesthetics to the look and feel of the final product.

Each creation conveys a message or a feeling about the product, system or environment. The aesthetics of the design is determined by the arrangement and use of the elements and principles of design, as discussed in the following chapter. If a designer is unhappy with the aesthetics of a design they can alter these by changing one or more of the elements and/or principles of design.

2.1.5 Finance

Finance will impact on designing and producing as a budget is a necessary part of all projects. If there is a client they, guided by the designer, will set the entire amount to be spent. If there is not a client, the designer will need to set a budget, both for income and expenditure. This budget is based on research and the accurate costing of every aspect of the project.

When planning the budget for the project, the designer will need to consider the weekly and total amounts to be spent, additional funding supplied by business partners and sponsors, donated materials and free labour supplied by friends and relatives, and financial rebates from businesses, government and any sponsorship deals.

When considering expenditure, the designer will need to research the actual cost of the time and labour involved in creating, modelling and manufacturing the design, and the costs of the materials, tools and manufacture. Also to be included are the costs of fuels such as electricity, insurance and consumable items. Designers also need to consider their choice of materials. Sustainable or 'green' materials and products are often more costly. This means that the designer's values and ethics will impact on the cost of the project.

The budget should balance the income and expenditure and the total amount spent should be in line with a reasonable selling price for the product and a profit for both the designer and manufacturer.

2.1.6 Ergonomics

Ergonomics is about 'fitting work to people'. It is the process of designing or arranging workplaces, products and systems so that they 'fit' the people who use them. Ergonomics comes into everything which involves people. Work and leisure systems should embody ergonomic principles if well designed. Ergonomists use the data and techniques of several disciplines including:

- anthropometry, which measures data from various body shapes and sizes.
- biomechanics, which examines muscles, levers, forces and strength.
- environmental physics, which examines noise, light, heat, cold and radiation.
- vibration body systems, including hearing and vision.
- sensations applied psychology, which examines skills and learning.
- social psychology data on groups, communication, learning and behaviours.

The aim of ergonomics is to develop a comfortable, safe and productive work system. This means designers must consider human abilities and limitations in the design process.

Figure 2.1 Ergonomics is an important factor when designing for human use.

2.1.7 Occupational health and safety

Technology continues to change our lives, and along with change comes the responsibility to use technology safely to avoid injury. Whenever we work in practical areas such as food laboratories, industrial workshops and textile facilities we are exposed to potentially hazardous situations. These situations must be monitored and measures put in place to prevent accidents occurring.

To protect people, governments have passed safety laws, and to ensure these are followed governments train and employ safety officers. Their task is to inspect workplaces to ensure that the occupational health and safety rules and regulations are in place. If not, fines and jail terms can result.

To protect workers and school students safety rules are made very clear. Your responsibility is to read, understand and follow safety rules, and seek clarification if you are not sure of anything. Be aware that safety is usually colour coded: red means 'stop' or 'danger', amber means 'caution' and green means 'go'. Air is light blue, steam is silver-grey and electrical cables are light orange.

Unsafe practice is inappropriate behaviour in a certain situation, such as not wearing personal protection equipment (PPE). Personal safety relies on safe practices and conditions and the use of appropriate safety gear. No matter how simple a task is you should always use the correct PPE and never take shortcuts.

An unsafe condition is a situation where an accident may occur. Examples include poor lighting or ventilation, an untidy work area, slippery floors, and so on.

Work Cover NSW manages the state's workplace safety, injury management and workers compensation systems. Their mission is to sustain a working partnership with the NSW community to achieve safe workplaces and an effective return to work for injured workers.

The Authority administers the:

- *Workers Compensation Act 1987*
- *Workplace Injury Management and Workers Compensation Act 1998*
- *Workplace Occupational Health and Safety Act 2000*
- *Factories, Shops and Industries Act 1962*
- *Dangerous Goods Act 1975*
- *Workers Compensation (Bushfire, Emergency & Rescue Services) Act 1987*
- and the Regulations and Codes of Practice under those Acts.

The objectives of the *Workplace Occupational Health and Safety Act* are:

- to protect workers and the public in the workplace against the improper actions of others
- to set workplace and safety standards
- to provide practical advice on how to reduce risks in the workplace
- to establish workplace health and safety committees
- to appoint safety officers within workplaces and safety inspectors to enforce the laws
- to create awareness about health and safety within the community.

The Act has two regulations which cover:

- administrative matters, such as preparing work plans and records, and
- standards for risk management, such as exposure to hazardous substances.

Should people fail to follow these regulations, there are fines and jail penalties.

WorkCover NSW works with industry, the workforce and insurers to:

- promote a culture of safety through public awareness programs, education and other community activities
- improve the performance of the workplace safety, injury management and workers compensation systems.

The purpose of the Workplace Health and Safety Committee is to provide a forum for people to discuss and address health and safety issues. To do this, regular meetings must be held. The Act requires that a workplace health and safety officer be appointed and trained by management. The officer's duties include:

- inspecting the workplace to identify unsafe practices and conditions.
- ensuring others observe health and safety standards.
- investigating and reporting all accidents and injuries.
- advising the employer on related issues.

New laws are in place to protect workers. The *Occupational Health and Safety Act 2000* and *Occupational Health and Safety Act Regulations 2001* are written in plain English and aim to improve workplace safety.

The *Workplace Injury Management and Workers Compensation Act 1998* focuses on the management, prevention and administration of workplace injuries. The emphasis is on the safe, timely and permanent return to work of injured workers.

Material Safety Data Sheets (MSDS) are used for guidance in the safe handling and storage of hazardous materials. It also states the necessary PPE and includes the following information:

- the product name
- the chemical names of certain ingredients, including generic names
- the chemical and physical properties
- health hazard information, including toxicity, effects, first aid and advice to the medical profession
- precautions for safe use, such as exposure limits, ventilation, personal protection and flammability

- instructions for safe handling, storage and transportation, plus information on what to do in the event of spills and disposal
- the manufacturer's name, Australian address and telephone number.

In the event of an accident, the following procedures should be followed:

- remove the danger, for example, switch off the power
- signal for help
- apply first aid if you know the correct procedure
- assist as directed by qualified persons
- evacuate the area if necessary.

2.1.8 Quality

With globalisation, the design, manufacturing and consumption of products can be separated by large distances. Products are often designed in one company and produced in another. Although the product is shaped in the manufacturing companies who produce the item, it is the design quality that is the key to product quality because design specifications have a major impact on a product's performance, quality and cost.

Techniques such as early supplier involvement and frequent communication are being used to integrate supplier's activities into product development. Therefore, product quality greatly depends on the quality of the design-manufacturing chain.

When designing your MDP it is important that you create a final design in the form of a production sketch that shows front, back and exploded views (if appropriate), with all features labelled and measurements given. The inclusion of this production sketch will ensure that during manufacture every aspect of the design can be reproduced to a high quality.

2.1.9 Short- and long-term environmental consequences

Designers need to consider the environmental consequences of every aspect of their creations because the planet needs to become more sustainable. The United Nations predicts that there will be nine billion people by 2050, and therefore:

- more resources will be diverted into consumption, leaving less for productivity
- already limited resources will be severely stretched
- demands for health and education will continue to grow.

Current manufacturing and consumption levels cannot continue as:

- we will run out of natural resources
- the pollution created is irreparably damaging the environment
- we are struggling to dispose of products at the end of their life.

Conventional design and construction methods have been linked to environmental damage, including depletion of natural resources, air and water pollution, toxic wastes and global warming. In the USA, buildings consume sixty-five per cent of the annual electricity consumption, twenty-five per cent of the timber harvest, and forty-two per cent of the transportable water. Buildings produce thirty per cent of annual greenhouse gas emissions and building materials account for forty per cent of landfill waste. It has been found that:

The richest 20% of the world's population:	The poorest 20% of the world's population:
Consume 45% of all meat and fish	Consume 5% of all meat and fish
Consume 58% of total energy	Consume less than 4% of energy
Have 74% of all telephone lines	Have just 1.5% of telephone lines
Consume 84% of all paper	Consume 1.1% of paper
Own 87% of the world's vehicles	Own less than 1% of vehicles

One solution may be for the richest twenty per cent to be made more accountable for their global footprint. Ecological or carbon footprinting estimates how much land we will need in order to sustain our current lifestyles. To maintain our current lifestyle we need:

- 8.5 planets to absorb our carbon dioxide
- 6 planets worth of steel
- 3.5 planets to sustain cement supplies
- 3.5 planets to meet current timber demands.

If we are to continue with our current consuming lifestyles how will future generations survive on this planet? This issue can be partially resolved by clever designing. Visit the bestfootforward website at www.bestfootforward.com/footprintlife.htm to estimate your own carbon footprint by filling in the calculator reproduced below.

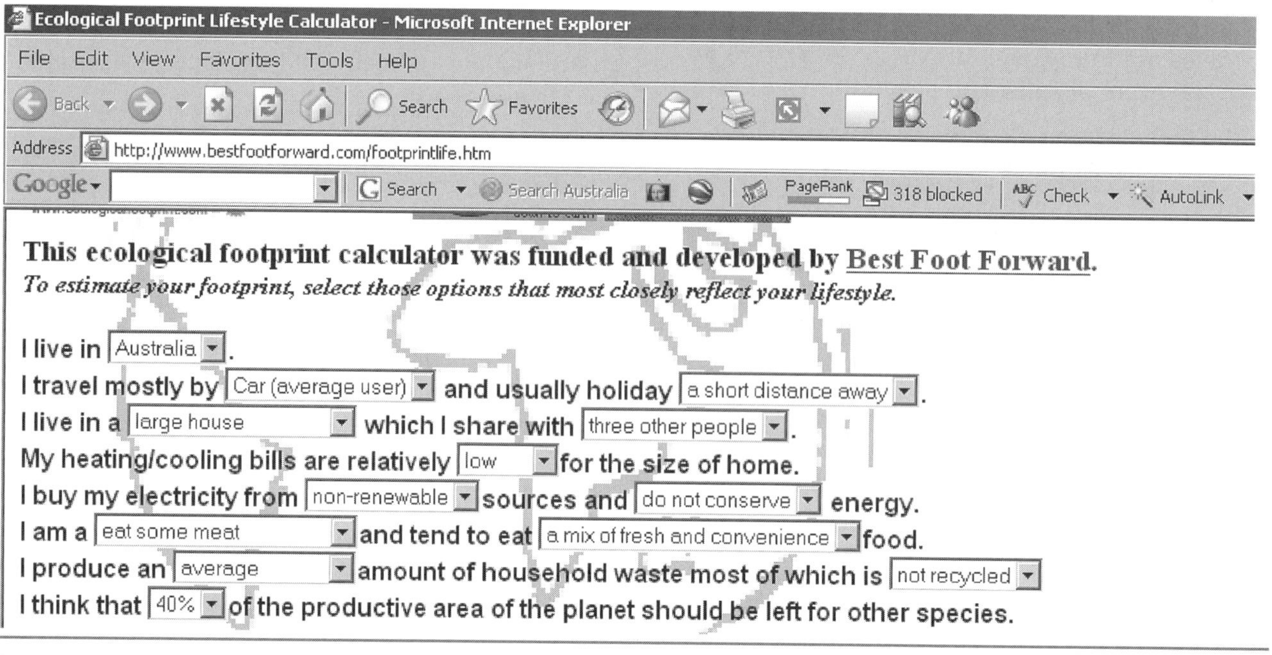

Ecological Footprint Lifestyle Calculator - Microsoft Internet Explorer

File Edit View Favorites Tools Help

Address http://www.bestfootforward.com/footprintlife.htm

This ecological footprint calculator was funded and developed by Best Foot Forward.

To estimate your footprint, select those options that most closely reflect your lifestyle.

I live in [Australia].

I travel mostly by [Car (average user)] and usually holiday [a short distance away].

I live in a [large house] which I share with [three other people].

My heating/cooling bills are relatively [low] for the size of home.

I buy my electricity from [non-renewable] sources and [do not conserve] energy.

I am a [eat some meat] and tend to eat [a mix of fresh and convenience] food.

I produce an [average] amount of household waste most of which is [not recycled]

I think that [40%] of the productive area of the planet should be left for other species.

Figure 2.2 Ecological footprint calculator
Source: Screenshot courtesy of Best Foot Forward Ltd (www.bestfootforward.com)

For further information on the environmental consequences of designs and designers refer to Chapter 5.

2.1.10 Obsolescence

Obsolescence is when a person, object or service is no longer wanted even though it may still be in good working order. Obsolescence frequently occurs because a superior replacement has become available. Some of the types of obsolescence are covered below.

Technical obsolescence

Technical obsolescence may occur when a new product supersedes and is found to be preferable to the old product. For example, the CD-ROM replaced the floppy disk because it allowed for greater storage capacity and speed and DVDs have replaced VHS tapes because they allowed for greater quality and multimedia functions.

Functional obsolescence

Some items become functionally obsolete when they do not function in the manner they did when created. This may be due to natural wear or some intervening act. For example, if new mobile phone technology is adopted and there is no longer a provider service based on the old technology, any mobile phone using that technology would be rendered obsolete. Products which naturally wear out or breakdown may become obsolete if replacement parts are no longer available, or when the cost of repairs exceeds the cost of a new item.

Planned obsolescence

This is the process of a product becoming obsolete after a certain period in a way that is planned by the manufacturer. Planned obsolescence has benefits for a producer because the consumer is under pressure to repurchase the item. This strategy generates long-term sales by reducing the time between repeat purchases. This is referred to as shortening the replacement cycle.

The use of planned obsolescence is complicated by related problems, such as competing technologies which expand functionality in newer products. Critics claim the process wastes resources and exploits customers. Supporters claim it drives technological advances and contributes to material well-being.

Planned obsolescence is encouraged by making repair costs comparable to replacement costs, or by refusing to provide service or parts. Planned obsolescence is common in batteries made to specifically fit a product. Many portable consumer electronics contain proprietary, often lithium-based batteries that are manufactured to specific sizes so when they run out they can only be replaced by the specially modified batteries.

Style obsolescence

Marketing may be driven primarily by aesthetic design. Products in which this is the case display a fashion cycle. By continually introducing new designs and retargeting or discontinuing others, a manufacturer can 'ride the fashion cycle'. The style changes are designed to make owners of the old model feel 'out of date', while some marketers go further by attempting to initiate fashions or fads.

Notification obsolescence

Some companies have developed a very sophisticated version of obsolescence in which the product informs the user when it is time to buy a replacement. Examples of this include water filters that display a replacement notice after a predefined time and disposable razors that have a strip that changes colour.

Postponement obsolescence

This refers to a situation where technological improvements are not introduced to a product even though they could be. One example is when an auto manufacturer develops a new feature for its cars but chooses not to implement that feature in the least expensive vehicles in the product line.

Obsolescence and durability

If designers expect a product to become obsolete they can design it to last for a specific lifetime. This is done through a technical process called value engineering. An example is home entertainment electronics which tend to be designed and built with moving components such as motors and gears that last until technical or stylistic innovations make them obsolete. These products could be built with higher-grade components but are not because it is felt that this imposes an unnecessary cost on the purchaser. Value engineering will reduce the cost of making the product and thereby lower the price to consumers.

2.1.11 Life cycle analysis (LCA)

Life cycle analysis has begun to be used to evaluate a city or region's future waste management options. The environmental assessment covers the environmental and resource impacts of alternative disposal processes.

The concept of conducting a detailed examination of the life cycle of a product or process is relatively recent and emerged in response to increased environmental awareness on the part of the general public, industry and governments. Taking as an example the case of a manufactured product, a life cycle analysis involves making detailed measurements at all stages of manufacture, from the obtaining of the raw materials used in production and distribution, through to the product's use, possible re-use or recycling, and eventual disposal.

Life cycle analyses enable a manufacturer to quantify how much energy and raw materials are used, and how much waste is generated at each stage of the product's life. Such a study would normally ignore second-generation impacts such as the energy required to fire the bricks used to build the kilns used to manufacture the product. Much of the worldwide focus has been on agreeing to the methods and boundaries used when making such analyses.

While carrying out an LCA is a lengthy and detailed exercise, the data collection stage is relatively uncomplicated as long as the boundary of the study has been clearly defined, the methodology rigorously applied, and reliable, high-quality data is available.

The second stage—life cycle assessment—is more difficult as it requires interpretation of the data and value judgements. For example, a life cycle inventory will reveal how much pulp, electricity and water is involved in producing a quantity of paper. Only by then assessing those statistics can a conclusion be reached about the product's overall environmental impact.

Some studies attempt to group the various impacts into clearly defined categories, for example, the possible impact on the ozone layer or the contribution to acid rain. Others go further still and try to add the grouped figures to arrive at a single 'score' for the product or process being evaluated. A great deal of work is currently being conducted on this aspect to arrive at a standardised method of interpreting the collected data.

Recycling introduces a real difficulty into the assessment calculations. In the case of materials such as steel and aluminium, which can technically be recycled an indefinite number of times, there is no longer a 'grave', the grave being the end disposal point of a product. In the case of paper it can be recycled through theoretically different types of collection schemes, resulting in cradle to cradle rather than cradle to grave.

All products have some impact on the environment but some use more resources, cause more pollution or generate more waste than others, so the aim is to identify those which are most harmful. Detailed analysis of the manufacturing process can also be an aid to identifying the use of scarce resources by showing where a more sustainable product could be substituted.

Despite the usefulness of LCA, in most situations it is impossible to prove that any one product or process is better in general terms than any other since many parameters cannot be simplified to the degree necessary to reach such a conclusion.

Key concepts and definitions

Design—a plan or creation that is conceived as a solution to a problem. It may involve sketching, researching, drawing up plans and/or construction.

Ergonomics—the aim of ergonomics is to develop a comfortable, safe and productive work system by considering human factors and data. It means designers must consider human abilities and limitations in the design process.

Evaluation—an assessment of the positive and negative attributes of design ideas, tools, materials and processes. It is an ongoing process.

Production is used in the process of construction, fabrication or assembly.

Useful websites

- www.baddesigns.com
- www.bestfootforward.com/footprintlife.htm
- www.cbmgroup.com.au/home.html
- www.curriculumsupport.education.nsw.gov.au/secondary/technology/7_10/technology/teaching_ideas/factors_affecting_design/index.htm
- www.dsol.com.au/

Classroom activities

1. For your current design project, describe in detail how each design factor has been considered and explain how you have used each factor to contribute to the success of the design.
2. Select a poor design from www.baddesigns.com and describe how each factor contributed to the failure of the design.

Key concept questions A pp. 196–197

1. Using your digital graphic design skills, create an A3 safety poster for your classroom that demonstrates how the Australian Workplace Health and Safety rules apply in the practical classroom at your school.
2. List the steps needed to construct your design project and beside each one suggest how the safety issues were addressed.
3. Complete the table below and on the following page, identifying and evaluating the aesthetic and functional properties.
4. Consider the mobile phone, where planned obsolescence is used, and describe how this concept has been incorporated into the product's development.

Product, system or environment	Graphic	Aesthetic properties	Functional properties
Sydney Harbour Bridge			

Product, system or environment	Graphic	Aesthetic properties	Functional properties
Sydney Opera House			
Backyard pool surrounds			
Formal gown			
Snowboard			
'Aqua' aftershave bottle			
Poached pears			

Sample Preliminary questions

Extended-response questions A p. 198

1. The lounge pictured below was created by leading Australian designer Marc Newson. It is now an icon and originals can be purchased through Marc's website in one-eighth scale for $1200. His client was a European art gallery owner who asked him to design a lounge chair that represents the new millennium.

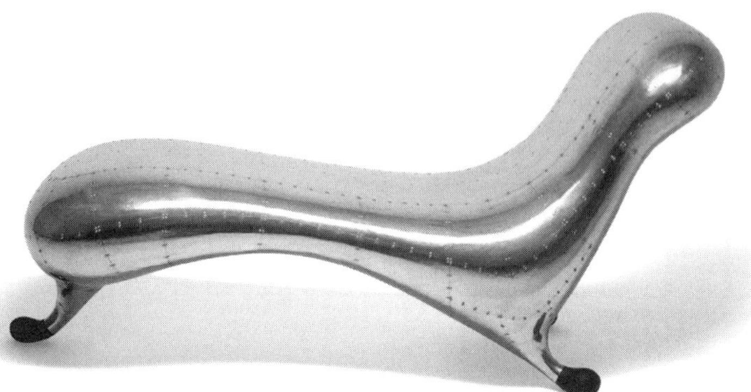

Source: Image supplied courtesy of Marc Newson Ltd. Lockheed Lounge, 1985, aluminium and fibreglass.

Explain how the following factors contributed to the success of the Lockheed lounge design:

- appropriateness of the design solution
- needs
- function
- aesthetics
- elements of design
- finance
- ergonomics
- occupational health and safety
- quality
- short- and long-term environmental consequences
- obsolescence
- life cycle analysis. (15 marks)

2. Describe each type of obsolescence and explain the role of obsolescence within contemporary design. (15 marks)

Additional factors affecting design

PRELIMINARY OUTCOMES

A student:	You learn about:	You learn to:
P1.1 examines design theory and practice, and considers the factors affecting designing and producing in design projects.	■ designing theory and practice including – the history of design – the elements of design – the principles of design.	■ apply design theory to practice.

The 'learn about' and 'learn to' components above are not a part of the Design and Technology Stage 6 syllabus. However, a basic understanding of the history, elements and principles of design is essential in order to gain a deeper understanding of design theory and this is why this additional information has been included in the study guide.

3.1 Design theory and practice

Design is not about decoration or aesthetics: it is a concept, a process used to solve problems. Design is about finding the best solution to an identified problem. As with any newly acquired skill, design ability will improve with consistent use.

'Design' is a term that lays a foundation for the developing of every idea to a possible solution. In this, designers think, evaluate and adjust their ideas in an interactive way. A plan is implemented and a design process is followed.

A designer works with both objective and subjective requirements. The objective requirements are technical and business requirements that allow for measurement and direct comparison, such as how much will it cost, what is the best material, when can it be finished by, and so on.

The subjective, creative side of design is the aesthetic element that relates to fashion, human behaviour and emotions, and cultural influences such as the meaning of symbols. The elements and principles of design help us to understand these factors.

Designers bring human and cultural values to business problems, values that sell products and services, create demand and inspire customer confidence and loyalty.

3.1.1 The history of design

The history of design can be traced through the development of technology and design styles that have taken over our world.

- Initially the arts and crafts movement inspired designers in Europe and North America who rejected the processes of new technologies and viewed the mass-produced products of the modern era as tasteless and lacking in quality.
- 1890–1920: in Paris the term *l'art nouveau* (new art) was used to describe a groundbreaking style. Art Nouveau was so influential that it set a trend in architecture and product design. It is characterised by flowing designs featuring leaves, vines and wavy lines. However, instead of completely rejecting the new industrial production methods and materials some designers did use them as an inspiration.
- 1920–1930: the Art Deco movement started in Europe and North America. It was the opposite of Art Nouveau and featured geometric shapes and lines and bright colours. The objects that were designed were generally mass-produced using new materials such as plastics.
- 1900–1920: a group of German designers and architects worked together to produce a new style that used machined products. Walter Gropius, an architect, led this group which became known as Bauhaus. Gropius and his followers developed products that worked well and did not rely on decoration to make them attractive. The idea behind Bauhaus can be summed up as 'form following function'.
- 1914–1918: during WWI women became more involved in the world of design and technology as many men were off fighting. After the war some of these women went on to become synonymous with good design. They include Gabrielle 'Coco' Chanel for fashion design, Clarice Cliff for ceramics, Betty Joel for furniture and textiles and Marianne Brandt for metalwork and architecture.

The diagram on page 29 outlines some of the major developments in the history of technology.

The elements and principles of design

'Beauty was not the intended outcome. Beauty was a natural by-product of craft diligently applied to serious things' (Peter London, 1989).

Most products, systems and environments in society are designed as solutions to identified needs. Designs are created to improve the quality of life. Good design creates a positive physical, emotional and psychological comfort and addresses both functional and aesthetic properties.

Great designers consider each and every element and principle of design before finalising their ideas. The elements and the principles are the building blocks that make a great design and are applied to both the functional and aesthetic properties.

Functional properties are aspects that are used to make the design function or suit a specific purpose. Words used to describe the function of a design include strong, comfortable, relaxing, sturdy, tactile, hard wearing, ergonomically friendly and durable.

Aesthetic properties are decorative details that could be removed from a design without affecting its function. Words used to describe the aesthetics of a design include pretty, tactile, smooth, bubbly, intricate, elaborate, complicated, simple and convoluted.

	Period	
Information Age Computers are commonplace and the Internet has 'reduced' the size of the world as it increasingly becomes a global village.	1990s	
	1950s	**Space Age** *Sputnik*, the first satellite, was launched in 1957.
Industrial Revolution Development of existing and new materials resulted in mechanisation and the development of new tools and machines that made factories more productive and less labour intensive.	1800s	
	1700s	**Steam Age** People now had a portable source of energy. Steam power was used for transport and in industry to run machinery to mass-produce products.
Iron Age Iron ore was found in abundance. Metal tools were made for crafts and farming. Coins were made and armies were well equipped with metal weapons.	1000 BCE	
	3500 BCE	**Bronze Age** People discovered copper and tin and combined these to form bronze. Cups and vases were made, as well as helmets, knives and shields.
Stone Age Neanderthal people used worked stone to cut meat for food and skins for clothing.	2.5 million BCE	

Figure 3.1 Stages in the development of technology

The elements and principles of design are arranged to improve the item. Giving consideration to and applying the elements and principles of design will create different forms and visual impacts by appealing to our senses and instincts.

3.1.2 The elements of design

The elements of design include line and direction, shape and size, texture, colour and value.

Line and direction

A line is a series of connected points. If there are two points the eye will make a connection and 'see' a line. This compulsion to connect parts is described as grouping or gestalt. A line is a mark that has a psychological impact that is determined by its direction and weight, and the variations in these. It can act as a symbolic language or it can communicate emotion through its character and direction. Lines exist in nature as a structural feature such as tree branches or as surface design such as striping on tigers, zebras and seashells.

Line defines direction as the type of line used will suggest movement, leading your eye in the direction it is travelling. Lines also divide spaces, outline objects and express emotions and feelings. Lines are significantly longer than they are wide. They can be hard or soft, which can correspondingly imply rigidity or flexibility. A line defines the direction of a design as it causes the eye to follow it, leading it around the design. Vertical lines can create the illusion of height as they lead the eye up and down. They give a feeling of elegance. Horizontal lines can create the illusion of width and a fuller appearance as they lead the eye from one side to another. However, the width of the line can affect these illusions. For example, thick vertical lines make a figure appear wider than thin vertical lines.

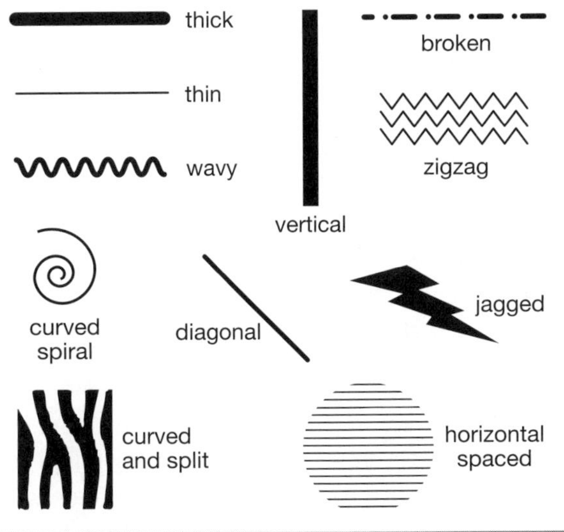

Figure 3.2 Types of lines

Line can create optical illusions. Your choice of borders and headings in design folios will involve selecting fonts that will contribute to the design and layout of the page. If you have selected images that use curves and circles then the fonts selected for the headings on the page should be of a similar style.

These lines are used for headings and support the layout of pages with straight line styles and graphics. Lines include borders and rules. They can be horizontal or vertical and help delineate the spaces around elements on folio and web pages. Line work increases the readability of a design.

Graphs are another linear device and are used to communicate quantitative information and relationships in a visual way. In your design folio you may include graphs to present results from experiments, questionnaires or surveys. The lines used in the layout of the page should reinforce the lines used in the graph.

Lines also communicate emotion through their character and direction. The psychological and emotional association with straight lines is that of a powerful, stable, masculine and dignified effect. Diagonal lines tend to create interesting, dramatic, dynamic, restless and indecisive feelings but they may unite contrasting directions or activity. Thin lines are seen as delicate, calm, fragile, subtle or sometimes weak, while thick lines are perceived as forceful, assertive, masculine and confident. A longer line is continuous, graceful, flowing or smooth while short lines are staccato, dynamic, efficient, abrupt or busy.

When designing an object consider the feelings that you are trying to convey. For example, in the design of a female outfit suitable for wearing in an office the feelings evoked should be respect, professionalism and efficiency. The outfit should say that the wearer is dressed for success.

Horizontal lines suggest a feeling of rest. Objects parallel to the earth are at rest in relation to gravity. Compositions in which horizontal lines dominate tend to be quiet and restful in feeling. Frank Lloyd Wright's architectural style uses strong horizontal elements which stress the relationship of the structure to the land.

Vertical lines communicate a feeling of loftiness and spirituality. Erect lines seem to extend upwards beyond human reach, towards the sky. They often dominate public architecture, from cathedrals to corporate headquarters. Extended perpendicular lines suggest an overpowering grandeur beyond ordinary human measure.

Diagonal lines suggest a feeling of movement or direction. Since objects in a diagonal position are unstable in relation to gravity, being neither vertical or horizontal, they are either about to fall or are already in motion. If a feeling of movement or speed is desired or a feeling of activity, diagonal lines can be used.

Figure 3.3 The use of vertical and horizontal lines

Horizontal and vertical lines in combination communicate stability and solidity. Square and rectilinear forms stagnate in relation to gravity, and are not likely to tip over. This stability suggests permanence, reliability and safety. Deep, acute curves, on the other hand, suggest confusion, turbulence, even frenzy, as in the violence of waves in a storm, the chaos of a tangled thread or the turmoil of lines suggested by the forms of a crowd. Curved lines do vary in meaning, however. Soft, shallow curves suggest comfort, safety, familiarity and relaxation. They recall the curves of the human body and therefore have a pleasing, sensual quality.

Shape and size

The second element of design is shape and size. Shape is created through a closed line and is generally two-dimensional. When depth is added it becomes 3-D. These shapes are known as form.

Silhouette refers to the shape of a 3-D design once completed while space refers to the area within a shape. Shapes make up any enclosed contour in the design. Geometric forms are those which correspond to named regular shapes, such as squares, rectangles, circles, cubes, spheres, cones and other regular forms. Architecture, such as these examples by Frank Lloyd Wright, is usually composed of constructed geometric forms.

Figure 3.4 Falling Water

Figure 3.5 Guggenheim Museum

Not all objects are geometric: many designed forms have irregular contours. Nor are all naturally occurring objects organic: snowflakes and soap bubbles are among the many geometric forms found in nature. The common terms used in relation to form and shape in composition reflect the kinds of representations the forms have. If we can recognise everyday objects and environments we refer to them as being realistic or naturalistic. However, if the images are difficult or impossible to identify in terms of our daily life we refer to them as abstract. Generally, abstractions are 'abstracted' or derived from realistic images that in some cases have been distorted.

Different shapes create different sensory responses. Sharp angles and diagonal lines appear unstable, creating the illusion of movement and excitement. Soft, curved shapes are intriguing and create the illusion of femininity. Regular, strong shapes with straight lines are stable, creating the illusion of confidence and power. In a motorbike the feeling of power is instantly conveyed with the swept back lines and strong shapes, even before the machine is started up.

Space creates psychological responses. Large simple areas of space suggest calmness and confidence whereas small, complex shapes and spaces are busy and interesting but may be confusing and cause tension. If space is too crowed the eye becomes tired and distracted.

Texture

Texture refers to the quality and characteristics of the surface of a design. Texture is used to add interest, dimension and variety. Texture is most commonly associated with touch (tactile) characteristics but can also refer to the visual characteristics of what we see.

Visual texture relates to when a design appears to exhibit a texture but this is simply an illusion. For example, a fabric with a tiger skin print would appear to feel like fur but instead may be smooth. Texture on web pages is all visual as you cannot feel it, but you can use an image of natural or artificial textures to get the textured effect.

Texture may be described as rough, smooth, shiny, bumpy, fuzzy, prickly, stiff and clingy. Tactile texture relates to the actual feel and physical characteristics of the design. 'Hand' or 'handle' is a term used to refer to the feel of a fabric. Terms used to describe the tactile hand may include coarse, soft, stiff, smooth and bumpy.

Texture has an impact on each design as it contributes to the psychological and emotional feelings the item will evoke. Textures can also be used in fashion to both promote positive attributes and hide faults. The rules are that coarse and bulky textures have an enlarging effect while bold checks, wide wales on corduroy, thick piles on furs and fluffy surfaces conceal the body contours, making it appear larger.

Bulky fabrics can therefore be used to disguise the body shape while lightweight, flowing fabrics cling and are therefore more revealing. Dull, matt textures are the most slimming as they absorb the light. Shiny surfaces, on the other hand, reflect the light, causing it to 'advance' and therefore make the design appear larger. The more complex the texture the simpler the design should be otherwise the detail of the pattern is lost.

Colour

Colour is one of the most important and exciting elements of design as it is one of the first things to be noticed. The three components of colour are hue, intensity and value.

Hue refers to the name given to a colour. There are three groups of colour hues:

1. Primary colours: these cannot be created/mixed from other colours. Primary colours can be mixed in different combinations to create other hues. They are red, yellow and blue.
2. Secondary colours: these are created when equal quantities of any two primary colours are mixed. They include green, purple, and orange.
3. Tertiary colours: these are created by mixing a primary colour with a neighbouring secondary colour, or by mixing two secondary colours together. The tertiary colours include red-orange, yellow-orange, yellow-green, blue-green, blue-violet and red-violet.

A colour wheel shows the relationship between primary, secondary and tertiary colours. For a colour wheel image, refer to the inside back cover.

Warm colours are red, yellow, orange, yellow-orange, red-orange and violet-red. We associate warm colours with heat or the warmth of the sun. Warm colours appear to advance because of the predominance of yellow. This means that they attract more attention and therefore create an enlarging effect.

Cool colours are violet, violet-blue, blue, blue-green, green and yellow-green. These colours suggest cool sensations and we associate them with things such as ice, water and the sky. They appear to recede because of the predominance of blue. This means that objects and shapes appear smaller and more distant.

Intensity refers to the saturation or level of chroma of a colour. The more grey a colour has in it, the less chroma it has. The intensity of a colour refers to its strength and purity. For example, two colours could be the same hue with the same value yet still have different colour strengths, such as a bright, clear red and a dull terracotta red.

Colour schemes represent the relationships between colours in a design. Monochromatic colour schemes use the shades and tints of just one colour to create a soothing effect. Harmonious or analogous colour schemes use colours that lie next to each other on the colour wheel. These are related as they have been mixed from a common hue. Yellow and orange are harmonious as they both contain yellow pigment. This scheme produces a rich, peaceful, pleasing effect.

Complementary colour schemes use colours opposite each other on the wheel, such as a warm and a cool colour. This produces a strong, bright and contrasting effect. Contrasting colour schemes have no hue in common, such as red and green.

Split complementary colour schemes contain two colours adjacent to each other on the colour wheel and one opposite. This creates a contrasting but more harmonious effect. Triad colour schemes use three colours evenly spaced on the colour wheel, giving a vibrant, strong and contrasting effect. Achromatic colour schemes use only black, white and grey tones.

The psychological and emotional effects of colour

The choice of colour in design reflects personality and taste. Colour is often associated with gender, age, social status, religion, culture and customs such as marriage. Colour can be used to evoke or create emotional and psychological responses. In plays and scenes in films colour suggests and evokes feelings. Green can suggest jealousy, red stands for danger and warning or romance and love, while yellow can suggest happiness.

In many cultures colour is associated with gender: in Western societies blue implies male and pink implies female. This colour association originated from cultural and spiritual connections. In modern society people's clothing is breaking down these associations as people wear colours that appeal to them, but the association or even stigma still remains. Colours can also suggest age: bright, pastel shades suggest youth while darker and more solid colours are associated with more mature ages.

The effects of colour

Optical illusions created by of colour include bright colours make figures appear larger while dark and dull colours make the figure appear smaller.

Strong dark colours look heavy; cool light colours can look lighter in weight; white can be used to unite colours; while black can be used to separate colours; shiny smooth textures enhance colour, making them appear brighter, livelier and larger than dull, rough surfaces.

Colour is the one design element that most designers are acutely aware of. However, colour is not a required element of design and a good approach is to create the design without colour first then add it to enhance the object.

Colour	Emotional	Physical	Visual	Over-use
White	Stimulating	Causes glare	Gives maximum feeling of space	Clinical
Off white	Suggests luxury, delicacy	Spacious	Gives background display for other colours	Monotonous
Black	Dramatic, sophisticated	Depressing	Adds spirit and interest	Dark and overpowering
Grey	Neither calms nor stimulates	Neutral	Neutral	Dull, uninteresting, unimaginative
Red	Exhilarating, exciting, aggressive and bold	Increases muscular tension, heat	Advancing, warm, bright	Overpowering, irritating
Yellow	Cheerful	Stimulating	High luminosoity, warm, sunny	Dominates other colours
Blue	Subdued, tranquilising, expresses distance	Decreases muscular tension	Cool, soft, atmospheric	Cold, empty
Green	Restful, refreshing, pleasant	Beneficial to eye nerves	Appears warm with yellow and cool with blue	Monotonous
Purple	Impressive, stately, dignified	Disturbs eye focus	Austere	Heavy, oppressive

Figure 3.6 Physical and emotional applications of colour

Value

Value, also referred to as 'tone', relates to the lightness or darkness of a colour or hue. These are referred to as shades or tints. A hue with black added is a shade as it has a darker value. For example, burgundy is a shade of red and a hue with white added is a tint as is has a lighter value, as in pink being a tint of red. The value spectrum (see inside back cover) shows the shades and tints of each hue.

The value spectrum is an important dimension of colour because the eye responds immediately to patterns created by the interplay of light and dark and the moods this interplay creates. Variations in value can emphasise, conceal, produce movement, add depth and volume and evoke emotion. The value scale is a graduated scale of nine tones ranging from white to black.

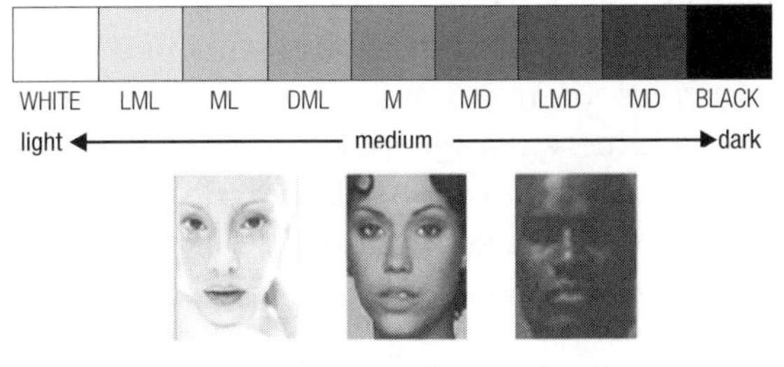

Figure 3.7 Value scale

When used in design value evokes an emotional and psychological response. Colours from the same value scale blend together well to create a soothing effect.

This form of colour scheme is also more slimming. Values that are similar (close together on the value scale) blend easily, creating a harmonious and soothing effect. These are harmonious values. Values that are different (far apart on the value scale) provide a contrast, which can be used to provide emphasis and add interest in a design. These are contrasting values.

3.1.3 The principles of design

The principles of design are used to arrange the elements, ensuring that all are integrated together to synthesise a new design.

The principles of design are defined as the ways in which the elements of design are applied. They are the tools designers apply to the elements. Visual design affects us emotionally and psychologically because our senses and instincts react to what we see. Therefore each design, each variable combination of elements and principles, can cause us to respond differently.

The designer's purpose drives the decisions made to achieve harmony between all the elements. This is achieved through the sensitive balance of variety and unity. The principles of design are proportion, balance, rhythm, emphasis or dominance, contrast, harmony, repetition and unity.

Proportion

This refers to the way in which all the parts of a design relate to each other individually and to the design as a whole. It is a comparison of sizes, shapes, amounts or parts. This is done by measuring with the eye. It is a synthesising principle as the process of comparison unites the separate design component pieces. Proportion refers to the relative size and scale of the various elements in a design. The success of the design is dependent on the relationship between the object and the parts of a whole. Proportion can be achieved decoratively through the application of features such as line, shape, colour and texture.

Proportion is structurally determined by the placement and positioning of the design elements. Every aspect of a design, whether structural or decorative, will alter the proportion of the design parts and therefore the design as a whole. Scale is an important aspect of proportion that deals specifically with size. The scale of a design is more effective when all elements are consistent. When the scale is consistent harmony, unity and balance can also be achieved.

A consistent scale is as important in an architectural drawing as it is in a pattern in textiles. Scale plays an important role in determining the success of a design. Scale also applies to the human proportion when clothing or accessories are concerned as these must be balanced to scale. It is also important in solving problems related to ergonomics.

Balance

Balance occurs when the visual weights (shapes, textures, colours, lines and spaces) of the parts of a design are equally distributed so as to create equilibrium. Balance gives a design strength and stability. The position of a feature will affect its visual weight. There are three types of balance:

1. Symmetrical balance is when the elements used are equal on either side of an axis, creating a predictable, serene and stable effect through the even distribution of the design's weight. Horizontal balance occurs when the left and right sides of a central axis are equal while vertical balance is created when the upper and lower parts of a design are equal.
2. Asymmetrical balance is when the elements used on either side of an axis are unequal or not distributed evenly. This creates the illusion that the design is 'heavier' on one side.
3. Radial balance is when the elements used radiate out evenly from a central point.

Rhythm

Rhythm refers to an organised movement, a visual 'beat' that stimulates the eye to move around the design. It gives a sense of flow and can become stronger when elements such as line, shape, colour and texture are repeated. Rhythm is achieved when the elements of a design are arranged in a predictable pattern.

A regular arrangement of the elements of design can create depth. The repetition of colours, textures, shapes and lines will emphasise a certain beat. Organised movement creates a mood and feeling in the design, for example, a flowing skirt with ruffles creates a soft, subtle feeling. There are three types of rhythm:

1. Regular rhythm refers to the repetition of an identical pattern.
2. Graduated rhythm refers to when the pattern increases or decreases.
3. Random rhythm refers to irregular intervals throughout a design.

Rhythm can be applied to create a mood or feeling within a design. For example, smooth undulating lines are indicative of gentle waves and therefore create a peaceful and calming effect while sharp jagged lines are forceful and exciting.

Rhythm can be achieved structurally and decoratively. Examples of structural rhythm include flowing hemlines, a curved stairway, and the curved seam lines on the Sydney Opera House which break the architectural sails into soft rhythmic panels. The straight lines and sharp angles of the Sydney Harbour Bridge create a more abrupt and staccato rhythm.

Emphasis or dominance

Emphasis, also known as dominance, creates a focal point in a design and is created by placement, contrast or isolation. This is the aspect of the design that first catches the viewer's attention. It may be an area of decoration or an aspect of the design structure that stands out from the rest. When there is no dominant feature in a design the eye may become bored and restless.

A variety of different methods can be used to create emphasis including manipulating elements to create focal points or using contrasting colours, lines and shapes to highlight features. Any element can be used to achieve emphasis. However, when there are several features of equal visual strength the eye can become distracted as various parts of the design are demanding attention. This means that the eye jumps from point to point attempting to find the most important feature. Emphasis is therefore a highlighting quality and as such can be used to draw attention away from any flaws in the design.

Contrast

Contrast exists when there is an unexpected conflict, tension or change in the visual elements of the design. It is an unexpected 'twist' in a design. If a design has too much contrast it can lose cohesion. The intense competition between parts creates an unsettling, unsatisfying and incomplete feeling. Contrast should not distract from the overall effect of a design and instead should be used to make it appear complete.

Contrast is a highlighting principle and the extreme differences it creates can magnify and draw attention to the opposing qualities of the elements concerned. It can be bold and exciting or aggressive and disturbing and therefore the placement of contrasting features is important.

Contrast can be achieved both structurally and decoratively. In clothing, it is achieved structurally by the placement of horizontal and vertical edges, seams, yokes, darts, waistlines and hemlines. It is achieved decoratively through trims, patterns and the application of any of the elements of colour, texture, line and shape.

Figure 3.8 Decorative contrast

Figure 3.9 Structural contrast

Harmony

Harmony occurs when one or more qualities of a design are alike. This repetition of similar features creates the feeling of agreement and consistency within a design. For a design to achieve harmony the two aspects of structure and decoration must be in accordance with each other.

Repetition

Repetition is directional as a repeated design feature encourages the eye to follow it. This movement can be used to emphasise the links between common design features. For example, repetition can emphasise width if it is used vertically. Repetition occurs when a line, shape, value, colour or texture is repeated more than once. It emphasises the psychological, physical and visual qualities of the element being repeated.

Regular repetition is when all aspects of the repeat are the same. This may result in a design that is tedious and boring as it lacks variety.

Irregular repetition is when there is a slight variation in the feature being repeated. This may cause a feeling of uncertainty but it will result in a design that is more interesting as the reduction of the repetition makes it less predictable.

Unity

Unity is the culmination or final point of a design and is achieved when every component, detail, feature and element and principle of design supports the central concept and overall quality, creating a sense of completeness and oneness. The structure, function and decoration all interact with each other to unite a design. All designs should strive to achieve unity. Harmony can also be created when a design looks like it 'belongs' together and is unified.

Unity is static when based on regular, harmonious features, creating a feeling that the design is solid and passive. Unity is dynamic when fluid, lively features are combined with irregular shapes, creating the illusion that the design is active and constantly moving.

Key concepts and definitions

Aesthetic properties—decorative details that could be removed from a design without affecting the item's function.

Elements of design—the basic parts of the design that work together to create a specific look, image or feel.

Functional properties are used to make the design 'work' or to suit a specific purpose.

Principles of design—general principles that can be used to create a psychological feel to a design. This is done by varying and contrasting the elements of design within each principle.

Useful websites

- bokardo.com/archives/five-principles-to-design-by
- desktoppub.about.com/od/elements/Elements_of_Design.htm
- jdh.oxfordjournals.org/
- www.designhistory.org
- www.johnlovett.com/test.htm
- www.swinburne.edu.au/design/tutorials/design/design/

Classroom activities

1. Select a product within your classroom such as a table, chair or cupboard and describe its function, psychological feel and appeal.
2. Discuss how each element of design was used to create the overall 'feel' of this product.
3. Using the same product explain which elements were used and how they supported the design's psychological aspect.
4. Evaluate this product in terms of whether the design was successful. Consider the brief, the elements used, the principles selected and the final solution. Note that if any of the elements contradict each other the design may not work.

Key concept questions

A p. 199

Figure 3.10 Architecture

Figure 3.11 Furniture

Figure 3.12 Fashion

1. Visual effects are created by line and direction. Analyse each of the three designs opposite. Describe the psychological and emotional response to the designs and then explain how line was used to create these feelings.
2. Using the colour wheel on the inside back cover select two different harmonious colour schemes such as yellow-orange and green-blue and apply a colour scheme to either a garment or a surfboard. What effect do harmonious colour combinations have? Then select two contrasting colours and apply them to another board. What effects do contrasting colour combinations have?

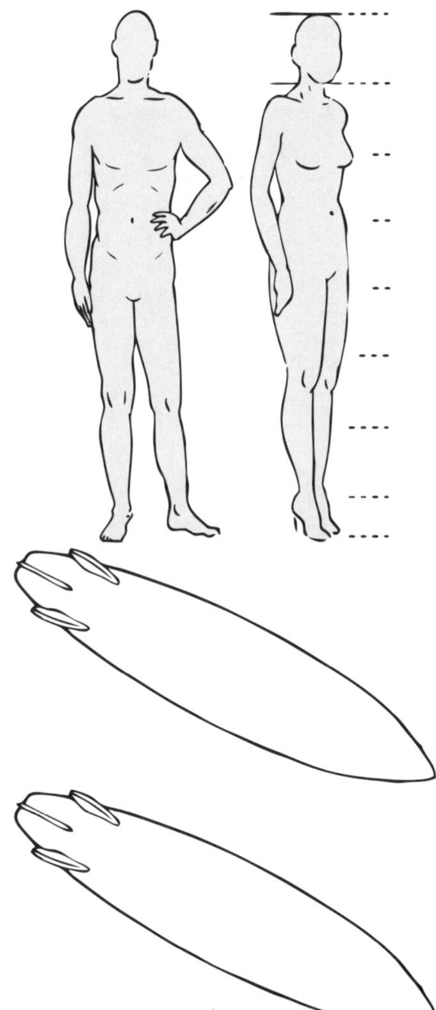

Figure 3.13

3. Analyse one of the designs below and discuss how the following elements and principles of design worked together to create unity:
   - line and direction
   - shape and size
   - texture
   - colour
   - proportion
   - balance
   - repetition
   - contrast
   - emphasis
   - rhythm.

   Explain why unity is important in a design.

Figure 3.14 Fashion design

Figure 3.15 Dubai architecture

4. (a) Describe your design brief in your own words. Include your design situation, limitations and problem.
   (b) Complete a production sketch of your design including a front and back view. This sketch must be assessed, evaluated and adjusted after answering each question below.
   (c) Describe the psychological and emotional response that you hope your design will evoke from others.
   (d) Identify how each element of design listed below has been used in your sketches:
   - line
   - direction
   - shape
   - size
   - texture
   - colour
   - value
   - unity.

   (e) Does each element convey the same psychological response? If not, adapt the element that needs changing in the design and re-sketch the creation.
   (f) Consider each principle of design carefully and decide whether or not they also support the psychological response to your design.
   (g) What type of balance is being used? It is symmetrical, asymmetrical or radial? How do the elements work together to create this effect?
   (h) What is the focal point of your design? Where do you want people to look first? How did you use the elements to ensure that you achieved this?
   (i) How has repetition been used in any element to create the visual rhythm of your design?
   (j) Explain how contrast and/or harmony were used to unify this design. Make further adjustments to your sketch if necessary.

Sample Preliminary questions

Short-answer questions A p. 200

1. How are the elements and principles of design used to create good design? (2 marks)
2. List the terms used to apply the elements and principles of design to the functional and aesthetic properties. (2 marks)
3. List the elements and principles of design. (2 marks)

Design and production processes

PRELIMINARY OUTCOMES

A student:	You learn about:	You learn to:
P2.1 identifies design and production processes in domestic, community, industrial and commercial settings.	■ design and production processes in domestic, community, industrial and commercial settings. ■ technologies in industrial and commercial settings.	■ compare and contrast technologies and processes used in design projects to activities of design and production in industrial and commercial settings.
P6.1 investigates a range of manufacturing and production processes and relates these to aspects of design projects.	■ manufacturing and production processes. ■ selection of processes appropriate to a need. ■ development of appropriate skills and techniques.	■ account for practices undertaken in industrial and commercial settings. ■ demonstrate quality production skills in the development of design projects.

4.1 Design and production processes in domestic, community, industrial and commercial settings

4.1.1 Design and production processes in domestic settings

Industrial-based technologies vary in many ways depending on the methods and technologies used. The variations in the size of the industry, company ownership, its role in the market, the type of activity, the industry sector from which it emerges or the technology of the day all impact on the manufacturing methodology a company employs.

The cottage industry

The term 'cottage industry' is applied to an industry-based development started in humble beginnings, when people manufactured hand-crafted goods in their homes in a part-time arrangement. Cottage industries were very common prior to the Industrial Revolution, when a large proportion of the population were engaged in agricultural activities. It was a way for farming families to earn extra money during the winter months. Those involved in cottage industries became proficient in their work and, as they developed their skills and began specialising, became known as artisans. These small manufacturing activities involved the use of locally available natural resources and human skills. Initially farming families would take in extra sewing or make cloth to be sold to a larger retailer. The whole family was involved and each member became responsible for a particular process, depending on their skills level.

The machines used were generally those already found in the home and the goods were not mass-produced: each design had its own individual flair. Some current designers are striving for this in their work and some student MDP designs imitate a cottage industry in the way manufacturing and technology are sacrificed for time and efficiency.

The cottage industry could be characterised by saying it was disorganised in its nature, used only local raw materials, and used technology found within the home itself.

The commodities produced were basically consumable items based on function and need, and were manufactured through the use of traditional techniques. The cottage industry helped to absorb a huge amount of surplus labour within the rural economy. Later, factory-based manufacturing was able to exploit this labour force and could also use cost-effective technologies that allowed them to supply products at a lower price than the labour-intensive cottage industries.

Today the term 'cottage industry' describes a variety of home-based employment activities and there are still many local productions of traditional handicrafts being developed. People see this as an extended hobby and an opportunity for an alternative way of life. It is also recognised by governments in developing countries as a way to engage the underemployed rural population and contribute towards export earnings.

People have realised that they are again able to support themselves, and there are large numbers of cottage-style industries or 'micro services' available with current technology making it easier to work from home. This is generally known as telecommuting, meaning the industry can be maintained without leaving home. The tele-industry makes use of online companies such as eBay, Overstock, Yahoo and Epier. Entrepreneurial activities that involve computer-based technologies often do not require the heavy set-up costs of a shopfront to develop their services. Generally they are able to conduct their business through web-based applications. The business propels its company, via a web page, and takes advantage of associated links.

4.1.2 Design and production processes in community settings

Small businesses or 'backyard industries' have always provided niche sales to customers. It may be the nature of the article, the uniqueness of the design or the sheer pleasure of owning the item. An abundance of design, creativeness and ingenuity has been borne out of Australian backyards with the most iconic being Mervyn Victor Richardson's Victa lawnmower.

There are many interesting and varied technologies involved in backyard production. In some cases the technology and processes may be crude and simplistic while in others advanced technological equipment is employed.

Most cottage designers start from modest

surroundings and if successful develop their designs into flourishing enterprises. This is largely due to improvements in design development and the processes, materials and technology used, plus a great deal of focus, patience and determination, and sometimes sheer luck. Many backyard industries turn what was a pastime into a full-time occupation. With this can come the employment of additional staff to help meet production commitments. What follows is often relocation into a factory space, an expansion of the business and an increase in capital investment. The original design ideal can be left far behind and a modest design that started as a hand-made object can reach the realm of mass production and computer-controlled mechanisation.

Potential difficulties that designers may face include market reluctance prior to supply and quality being guaranteed, competition from others with lower labour and other costs, insufficient data relating to product trends, and a lack of comprehensive industry networking.

The outworker

Another home-based activity is that of the garment-making outworker. A manufacturing company will subcontract their garment making out to textile workers. While the government now has strict controls on the welfare of these home workers, this was not always the case. In the recent past outworkers were typically doing the same tasks as sweatshop workers of the 1970s and 1980s. They produced large numbers of articles and received minimal pay, no entitlements such as sick leave or holidays, and worked in conditions that could endanger their safety and health. Most outworkers were paid 'cash in hand' and before regulation of the industry were earning as little as two to three dollars per hour: this resulted in the need to work for longer periods of time. Outworkers generally opted for this form of employment because of low skill or language levels or the need to combine work with caring for children. All major garment manufacturers currently employing outworkers ensure that their workers are treated with a high degree of ethics. There are two acts that aim to protect workers: the *Occupational Health and Safety Act 1989*, which is known as the *Prevention Act,* and the *Accident Compensation Act 1985*.

4.1.3 Design and production processes in industrial settings

Comparing the processes of large and small businesses to the processes of a Design and Technology student will show many variations. Similarities include the fact that all are involved in the design and manufacturing of a product, system or environment, from raw materials to finished goods. Manufacturing on a small scale is design linked to a need or want that can be personal and unique, a community need, or a global problem. Manufacturing on a large scale is closely connected to engineering and industrial design where processes are designed based on performance, materials selection and production methods.

Improvements made in production methods and factory organisation include specialised mechanisation, precision manufacturing, inter-replacement of parts and carefully coordinated work sequences. While improvements were made in the production methods, the instant progress provided a problem with a surplus of parts at the factory's point of assembly. It was recognised at that stage that all factory organisations required the same organisational improvement from start to finish. This problem, first realised by Henry Ford, was also solved by him by streamlining assembly methods. This process is called 'repetitive flow' or 'series production' and was the first time large-scale production used standardised products in an assembly line. Ford used this technique in his own factories in the early twentieth century, notably with his Ford Model T.

Figure 4.1 Early mass production incorporated standardised techniques in the assembly line.

Production progressed to workers performing repetitive tasks in order to permit high rates of production. The units would move around the factory on tracks or conveyor belts set up as assembly lines. The initial idea was a success as it reduced the probabilities of human error and variations in the product. It reduced labour costs and increased the rate of production, enabling the manufacture of larger quantities of the same product at a lower cost than via traditional non-linear methods. It now seemed that a limitless number of virtually identical goods could be produced.

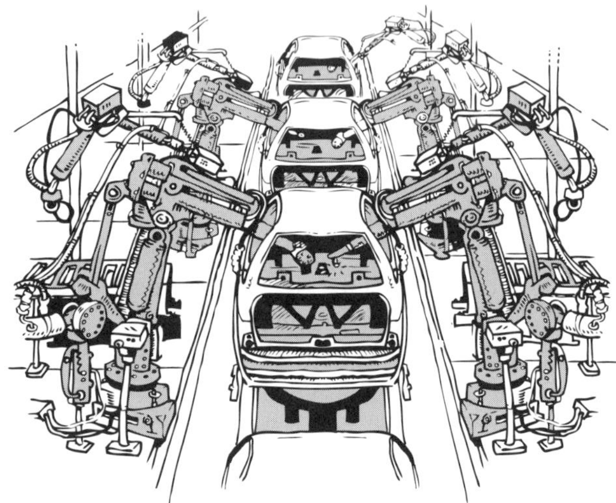

Figure 4.2 Modern car assembly plants require computer control on the assembly line.

As technology develops, so does the use of machinery in industry. The machinery that is needed to set up a mass-production line is expensive so companies need to ensure that production costs will be recovered. Technologies such as robotics are expensive to purchase and have high installation costs. Mass production is ideally suited to provide large numbers of the same product to meet the demands of a relatively homogenous population of consumers.

Due to the rigid production system, Henry Ford led his company into an economic and technical decline. Consumers were getting tired of 'the same', and over-capacity struck the automobile industry. General Motors, on the other hand, recognised that production needed to be flexible enough to cater for consumer demands. When their volume began exceeding demand they revised their approach to marketing the motor car. By introducing mechanical improvements and innovation to their products they were able to maintain interest and increase their market share.

General Motors perceived the individual nature of the consumer to the motor car and began emphasising style and superficial model changes as ideas for improvement. They moved forward by consciously aiming models at different income groups, which further increased their market share. This was to become known as 'flexible mass production'.

The evolution of traditional hand-operated machine tools into their modern counterparts of Computer Numerical Controlled (CNC) systems revolutionised the machines and parts industries. To maintain standards and specifications a certain level of control is required over the product, service and processes. Quality control practices examine and verify each characteristic of the product or service. Quality assurance (QA) is also included as a quality control measure. However, it has more to do with the process than the product as it ensures that the product or service is produced in the right way. If a problem is identified, the quality control team will stop production if necessary and correct the imbalance.

Another motor vehicle manufacturer renowned for its successes is Toyota. The company prides itself on quality control and continuous improvements resulting in what is known as 'lean manufacturing'.

Lean manufacturing is the production of goods using less of everything compared to traditional methods of mass production. It is borne from a Japanese concept of eliminating waste: the elimination of waste is the goal of lean manufacturing, and Toyota defined three types of waste: *muda* or non value-added work, *muri* or overburden and *mura* or unevenness.

Toyota has developed the Toyota Production System (TPS) in order to produce a new vehicle, by focusing on less waste, human effort, manufacturing space, investment in tools, inventory and engineering time.

4.1.4 Design and production processes in commercial settings

Manufacturing techniques vary widely depending on the product. The main types of production processes are the following.

- blow moulding: air pressure is used to make even, circular shapes. This process is used to create plastic bottles and containers.
- calendaring: a process of laminating fabrics with a polymer coating. It is used for shower curtains, tents and awnings.

- extrusion: where molten plastic is forced through a shaped die. It is used for plastic pipe tubing and sheets and other profile shapes.
- injection moulding is where molten polymer is injected into a mould. It is used for casings of most portable machinery, plastic bins, telephones, and so on.
- rotational moulding: also known as rotomoulding, it involves simultaneous heating and rotating of molten polymer in a closed mould. This process is used to make kayaks, grass catchers for lawn mowers and plastic drums.
- vacuum forming: a softened plastic sheet is drawn into an open mould by the atmospheric pressure of the vacuum. The process is commonly used in the manufacture of refrigerator linings, fast food containers and toy segments.
- compression moulding: plastic powder is placed under heat and pressure in a mould cavity which results in a set chemical change. It is used for electrical fittings, saucepan handles, and so on.
- rapid prototyping is the automatic construction of physical objects using additive manufacturing technologies employed from CAD operations. It is commonly known as 3-D printing.
- casting: molten metal is placed in a mould and allowed to solidify. It is used for metal game pieces and machine components.
- fabrication is a process of joining metals by using welding, brazing and soldering techniques.

Ways of organising these processes are covered below.

Custom production

This is one-off manufacturing such as hand-crafted or custom-developed production where a special order can be placed and a few people work on the product at any given time. Custom productions are common in the fashion industry, where one-off garments, usually from famous designers, are created and then worn, and thus seen, in up-market locations. It is also an opportunity for the designer to introduce and expose new design elements. Custom designs are often presented at fashion shows. Costuming departments generally use custom production to maintain the exacting detail required. The nature of the customising allows them to focus on the exactness and uniqueness of the detail as they do not need to consider mass-manufacturing processes. Custom production or 'customising' is seen in many different design fields including the areas of custom car and motorcycle design.

Batch-lot or job-lot production

When a production run with limited numbers is required, over a set period of time, the industry will use batch or job-lot production where requirements and volume are identified. Manufacturing machines are then set up to suit the conditions for the batch. Once the batch has been completed the machines are reconfigured for another batch requiring different manufacturing processes. Batch-lot production is popular in small industries as it helps keep costs down and meets the demand for the product.

Continuous production

When demand increases to the point where batch production cannot sustain the consumers, continuous production must be used. The production is developed to a point of steady supply and as an ongoing process. The products are generic in their design and cater for a mass market. Most designers at some point recognise that they are designing for this type of need. It is generally known as industrial design and the designer must consider the manufacturing method as a constraint. An assembly-line production is used when demand is high. Assembly-line production relies on people and/or machines specialising in a particular process as the product flows past. The fast, efficient production ultimately reduces the cost to consumers.

Total quality management (TQM)

TQM is a strategy aimed at ensuring all people in an organisation understand the importance of quality. Examining the way the consumer uses the product can lead to improvements in the product itself. It recognises the importance of all steps in the production process and is aimed at ensuring strong customer satisfaction, known as quality assurance. It is important that TQM is employed across all company departments. A large component of this process is the standardisation of products and processes. TQM has developed from a Japanese model known as Kaizen, which focuses on continuous improvement. The McDonalds Corporation is one of TQM's great success stories.

Value-based management (VBM)

VBM is an approach that ensures corporations use financial value as a performance indicator. This generally translates as maximising the company's value to its shareholders. VBM develops a strong corporate culture and monitors the performance of all employees. It is based on measuring a value over time to monitor its improvement. These measurements are known as 'value indicators'. In this way if an investment turns into a value it then becomes an asset to the corporation and a profit to the investors. To understand the concept, look at American office furniture company Herman Miller, which trains its workforce in VBM. Each month a non-finance employee must present the financial report to their colleagues to ensure that all understand this aspect of the company. In this way the importance of value is developed through the financial performance of the company.

Just-in-time management (JIT)

A just-in-time inventory system is designed to ensure that materials or supplies arrive at a facility only when they are needed in order to minimise storage and holding costs. The process relies on a series of signals, or 'Kanban', that trigger production processes. Kanban are usually 'tickets' but can be simple visual signals, such as the absence of a part on a shelf. The system relies on the company's approach to re-supplying, ordering and/or manufacturing of products to keep them in stock. The idea is used in most forms of mass-production and is lately being used by small companies advertising on eBay and trying to keep up with global demands.

4.2 Technologies in industrial and commercial settings

4.2.1 Technologies in industrial settings

Timber in an industrial setting

Some industrial timber production techniques can also be used in homes and schools with the machinery duplicated on a much smaller scale for the student or home 'handyman'. Differences include the high degree of mechanisation in industry and the degree of skill that accompanies such technology. As with any manufacturing industry, specific work practices and processes enable the production of timber products, usually within assembly plants. Designers and engineers are closely involved in the manufacturing process with the engineers interpreting the design and the most practical methods of producing it. Many designs need to be slightly adjusted to suit current design and engineering thinking, the raw materials, and the manufacturing methodology.

Production techniques for an individual

Timber technology at school, in a home industry or small factory can be classified as a custom production, which has the advantage of incorporating any design features the designer wishes. The only constraints are materials selection and joining methods. Completed projects can tend to look more like works of art than practical products. It is at this point where the designer and manufacturer are 'one' and function and aesthetics can meet. One problem is that as a one-off the item will be costly to produce. While many short-term production techniques help save time and money it is generally recognised that a one-off piece tends to have an expensive price-tag.

Production techniques for the timber industry

Mass-production methods generally cater to a broad group of consumers who require generic forms of furniture priced to suit their income. To meet this demand, raw materials are obtained and the furniture is manufactured and then stored or delivered to furniture outlets. The furniture is usually:

- ready to use: the product is assembled once it leaves the warehouse
- designed for assembly: the product can be easily transported in a 'flat pack' and assembled at home with 'knock-down fittings' and a set of instructions
- designed for disassembly: can be used when the piece is being moved or if the piece can be reused because of dwindling raw materials.

The generic styling, mass production and mass marketing of a limited range of designs allows for competitive pricing. While much furniture processed for the public is seen as generic in design it can be seasonally adjusted by superficially changing the style and fabric. Some furniture outlets will focus on style and individual design within part of their range but at the same time develop batch production items. This provides the timber industry with a 'best of both worlds' approach, allowing for savings to the consumer without losing the sentiment of strong individual design.

The technology

The size and capabilities of machines vary according to raw materials and product type.

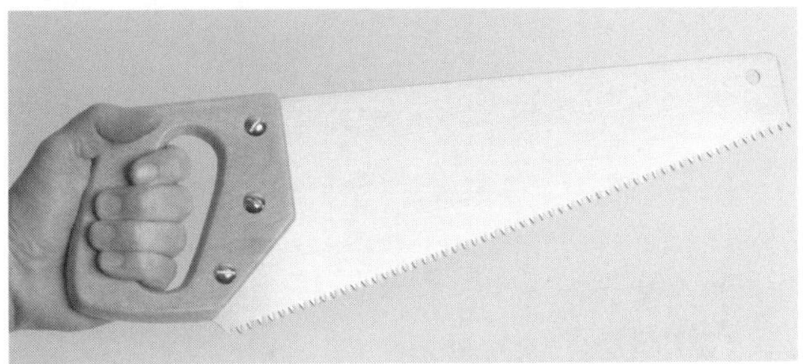

Figure 4.3 Hand tools require limited mechanisation and furniture may be costly to produce.

Figure 4.4 A high degree of mechanisation is used in industries producing generic furniture forms.

The workforce

The work of a furniture maker requires a variety of skills as, in addition to basic construction, there is the finishing and upholstery to be considered. Individual producers are multi-skilled while the advantage for large companies is the ability to recruit staff with specific skills that can be utilised along the production line.

Furniture making in an individual setting

All furniture design is needs-based and done at a personal level. While varied, the manufacturing process is completed at a lower level of technology to that used in large industry. Many individuals working from home-styled factories will use whatever tools and technology are available and can, within these limits, produce fine work that in some cases is achieved by replicating a scaled-down version of larger manufacturing processes. Where large manufacturers use CNC machines to produce identical pieces so too can the individual achieve this with the aid of tools such as jigs and gauges.

4.2.2 Technologies in commercial settings

Designing, raw materials handling, manufacturing, assembling, distribution and sales are features shared by industry and individuals. The manufacturing process is similar, with the only difference being the size of the operation. Manufacturing must be planned at all stages and on all scales if the process is to operate efficiently. The general steps in the manufacturing of products are:

1. Identify the customer need for the product through interpreted research and target market sampling.
2. Designing the product. The use of design tools, cognitive organisers, design team collaboration and acceptance of ideas are all involved. A variety of design techniques are used, from concept sketches with labelling to intricate CAD applications.
3. Concept sketches. Initial design development involving brainstorming techniques to identify the implications in product design and communicate these to the client or focus groups.
4. Product design identified and secured. Designing is collaborative, so all participants in the design team must be clear in their understanding and interpretation of the final design. There needs to be good communication and strong shared support for the end-process.
5. Working drawings. Details are finalised. Interpretation of the design at the construction stage, including the finalisation of the working drawings.

6. Prototype built and tested. The working model is trialled by user groups and evaluated in a range of different ways. Redesign can occur at this stage by fine-tuning any of the design features.
7. Marketing plan. Based upon the results of the prototype a marketing plan can be developed. It will include information based on investigating the needs of the target market.
8. Production sequence. Once the method of manufacturing has been decided on the production sequence can be finalised. This determines how each product is processed.
9. Materials. Materials are now ordered. They can include a range of raw materials and parts purchased from secondary industries.
10. Plant layout. A thorough final assessment of the factory would include checks on storage of raw materials and their processing and the containment of waste. The warehousing of the final product would impact upon the initial placement of machinery and the processes on the factory floor.
11. Tooling up. Engineers and technicians ensure that the machines and tools suit each application in the manufacturing process. In larger industries CAM (computer-aided manufacture) will be used. A greater investment in machinery will result in fewer employees. Mechanisation will increase the quality of the goods and generally maintain a division between highly skilled and unskilled operators on the factory floor.
12. Job run developed. Once job descriptions are drawn up, skilled workers are sought and others are trained. Workplace teams are organised and assist in factory communication, forming part of the TQM process.
13. Trial runs. Initial trial runs take place before 'real' production commences. This is to ensure the processes will operate without problems. It is also a time to evaluate the processes and make changes to improve speed and accuracy if necessary.
14. Full-scale production. The flow of production commences with all participants aware of their roles. Production rates are monitored, cross-checked and compared to statistics to ensure the industry is operating at full efficiency.
15. Quality control. The workforce is trained in quality assurance processes and measures are put in place to check the production process.
16. Warehousing and distribution. The completed products are stored in a warehouse and quantity is monitored. Packaging and distribution can be undertaken in order to meet demand.
17. Products sold. The product reaches the market where it is available to the consumer. Inventory rates are monitored to achieve a production versus sale balance.

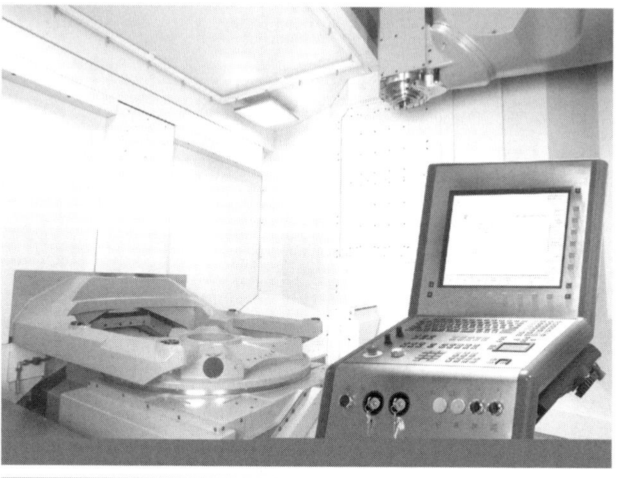

Figure 4.5 CAM (computer-aided manufacture) technologies are used in larger industries.

An individual producing a custom production needs to recognise that initial materials and tool costs are high. This is exactly what a company would experience when setting up. Fortunately for individuals they usually have some degree of flexibility but they still need to ensure that they stay on budget.

Companies and individuals have much in common and at times they think the same. All designing has an inspiration and develops into an action plan. All develop an industry commitment and move their ideas through to production. Both companies and individuals continually monitor the overall impact their designs have on the market and base their ideas on a series of decision-making processes.

Ethics and security

An important ethics concern for each designer/manufacturer is to look forward and determine a sustainable future. Any sustainable development involves responsible thinking about the environment, resulting in the selection of the right resources and technologies. The concerns relate to the responsible collection of raw materials and the elimination or control of waste products. Ethics is also related to the level of support and safety provided for each worker.

Whether working in an industry or as an individual, manufacturers need to recognise the importance of the design and consider the security of ideas, drawings, logos and products. This usually involves the use of security structures such as intellectual property (IP), copyright, trademarks and patents to protect ideas, manufacturing methods and overall designs. While most individuals may not be applying for idea protection, most understand how devastating it would be for a designer to find their styles and designs being copied or mass produced.

Project management tools

Management goals are generally recognised as getting the process completed more efficiently for a profit: in an individual setting one would use a management process to maintain a quality product through to the end. The TQM model may be used to accomplish this. Every small process is made to a 'best standard' and the range of processes to a 'best level', combining to form a high-quality product.

In industry a management process consists mainly of strategies to achieve goals. These can include strategic, tactical or operating plans. As an individual you may use all three plans when developing your MDP. You need to be strategic in the way you follow the syllabus and the marking guidelines, and tactical in reading the examiners' comments about the practical process and its marking. You must also produce an operating plan that allocates work time and deadlines for completing research and assignments. You may find you have to use all three plans at certain stages to manage your short- and long-term goals effectively.

Another management tool that you may need to use is Just-in-time management. Using this method will ensure that all your processes are completed by each required date. The type of management processes you are using are the action/time plan and the finance plan. The action/time plan predicts the length of time required to complete a project and is used to then plan the required actions. The finance plan predicts the project's cost to completion. As in industry, you need to keep to this plan so you do not overspend.

Documentation

Documentation is crucial for any manufacturing company and is equally crucial for your MDP as it communicates your design ideas. While it might seem obvious to you what you are planning and doing it is still essential to write it all down, including the inspirations that led to your MDP idea.

Communication is crucial in any collaborative team and designers must solve problems and make last-minute changes if necessary, which requires a good knowledge of the production process. All changes need to be documented for the knowledge of those involved in the project. Similarly, your folio should also document everything and it too will be 'checked'.

4.3 Manufacturing and production processes

4.3.1 Manufacturing and production

Textile manufacturing and production

Stitching techniques are similar in the home, school or industry. Textile workers use industrial machines which allow a wider range of techniques to be completed in safety and at a faster pace. Factories may contain over a hundred machines, including several computer-programmable units incorporating the latest technology. These range from heavy-duty stitching equipment to light-duty machines for finer work on lighter fabrics.

Individual production techniques

The production of an individual garment is known as a 'one-off' and is classified as custom production.

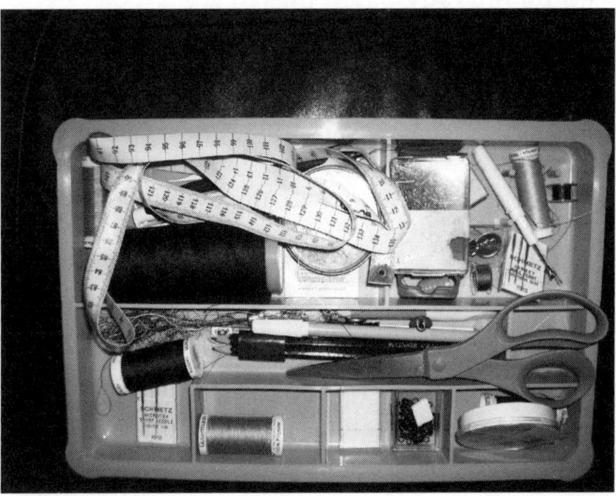

Figure 4.6 Textile technology at a personal level. While varied, the manufacturing process is completed at a lower level of technology.

With this technique the design can be corrected and directed at any time. The disadvantage is that it is costly to produce because raw materials and tools are purchased in small amounts and do not attract bulk purchase discounts. The processing is individual so no time-saving production methods can be employed. Some fashion companies will create one-off productions by well-known designers that are very expensive but attract celebrity clients who in turn provide publicity.

Production techniques for groups

Costuming for dance and theatre companies can employ either individual or mass production methods. The garments are generally themed costumes and while they are produced as a batch-lot production, individual characteristics can be added in a customised fashion. Whether job-lot or one-off production is used depends on the volume and nature of the required costumes and in some cases would combine the two, with costumes for the chorus being job-lots and those for the leading roles being one-offs.

Production techniques for the fashion industry

Mass and continuous production methods cater for the fashion industry where large numbers of items are produced. Manufacturers using JIT management are usually large enough to wait for an order and, with raw materials such as fabric on hand, can then go into production to swiftly meet the demand.

The technology

Large clothing companies which manufacture through mass-production will have computer-controlled systems where machines and robotics maintain the work processes. Patterns are generally scanned and digitised into the system or can be created directly on a screen through a CAD (computer-aided design) program where all functions, data and processes are easily identifiable through a special 'intuitive interface'.

'Smart' information about the fabric and its relationship to other pieces and styles is readily accessible through large data-storage capabilities. Programs can also detect user errors early, preventing losses of time and profit. The software allows for work on an unlimited number of pattern pieces at the same time and data can be modified directly and easily.

Materials purchased attract a discount, especially if larger amounts can be bought. Large companies place large orders and as a result receive a considerable discount. Technology is different for individual and commercial use. For example, individuals would use dressmaker's scissors but companies involved in batch production would use a cutting knife: an electric-powered machine that cuts through many layers of material at the same time, producing identical shapes.

The workforce

In the fashion industry the workforce is jointly responsible for the end result. Employees are encouraged to work together and this maintains quality control to some degree. Some textile workers are employed in large factories while others are outworkers who take the raw materials home.

Figure 4.7 The modern textile factory

4.4 Selection of processes appropriate to a need

When working through the design process it is crucial to have an idea of what direction to follow and what needs to done next. This will ensure strong management of the project, with clear goals and a well thought-out timeline. To ensure this process runs smoothly, list the step-by-step processes that need to be done. Following this process is made easier if it is tabulated and aligned with how each of these steps will be achieved, providing a guiding timeline to keep you on track. In your MDP this is seen as an action plan.

What has to be done	How each step will be achieved	Date to be completed
1. Research available materials.	■ Interview teacher ■ Talk to materials expert ■ Look on the Internet and at YouTube clips ■ Look in Yellow Pages ■ Study text books, journals and magazines	Week 3, Term 1

4.5 Development of appropriate skills and techniques

In every design project it is essential that all manufacturing and production be completed at the highest quality. In order to achieve this, all necessary skills must be used over a period of time to allow them to develop to a high level. This will help ensure the production of a superior product.

Key concepts and definitions

Computer Numerical Controlled (CNC) operation—automated machines operated by programmed commands encoded on a storage medium, as opposed to those that are manually controlled.

Cottage industry—the production and sale of goods at home as, for example, in the making of handicrafts by rural families. It may also involve any small-scale or loosely organised industry.

Custom productions—unique, one-off productions. They are very expensive and tend to attract a well-known designer to the design process.

Lean manufacturing—also known as lean production, this is a practice that considers the expenditure of resources for any goal other than the creation of value.

Mass production—the manufacture of goods in large quantities, often using standardised designs and assembly-line techniques.

Outworker—a person engaged to work away from the employer's commercial premises, for example, in their own home.

Telecommuting—home-based employment where communication with the workplace is by phone, fax and email.

Useful websites

- history-world.org/Industrial%20Intro.htm
- www.britannica.com/bps/additionalcontent/18/31941247/OCCUPATIONAL-HEALTH-AND-SAFETY-DUTIES-TO-PROTECT-OUTWORKERS-THE-FAILURE-OF-REGULATORY-INTERVENTION-AND-CALLS-FOR-REFORM
- www.homerun-business.com/cottage_industry
- www.referenceforbusiness.com/encyclopedia/Int-Jun/Japanese-Manufacturing-Techniques.html
- www.strategosinc.com/just_in_time.htm

Classroom activities

1. Use the following T-chart to compare your MDP production and management methods and technologies to that of an industry.

Used in industry	Used in MDP
Technological activities	Technological activities
Production methods	Production methods
Management methods	Management methods

2. Report on the techniques that you might use in marketing your MDP and explain how it would be produced on a commercial scale.
3. Describe the modern manufacturing industry as it relates to your current project.
4. Use a concept board and cut and paste images from the Internet that reflect your ideas on the form of individual and industrial settings.

5. (a) Visit the Australian furniture manufacturer Schiavello's website (www.schiavello.com/profile.htm) to identify the company's objectives. Select manufacturing (www.schiavello.com/manufacturing.htm) and summarise Schiavello's approach to manufacturing in regard to technology, management and the environment.
   (b) Visit the following site www.business.vic.gov.au/busvicwr/_assets/main/lib60040/10_schiavello_casestudy.pdf, page 4, and report on the problems the Schiavello Group face in a highly competitive and overcrowded market.

Key concept questions A pp. 200–201

1. Outline the popularity of the cottage industry.
2. Examine how the idea has developed into the modern-day 'cottage industry experience'.
3. Outline the home-based activity of the garment-making outworker.
4. Identify the two Acts that aim to protect the outworker.
5. Describe three manufacturing techniques used in industry.
6. Describe three management techniques used in industry.

Sample Preliminary questions

Extended-response question A p. 201

1. The work of a designer requires the use of a variety of production and management techniques.
   (a) Explain the production and management methods of the design and production of your MDP. (10 marks)
   (b) Predict how these methods would vary if your MDP was to be industrially produced. (5 marks)

Design for the environment

PRELIMINARY OUTCOMES

A student:	You learn about:	You learn to:
P2.2 explains the impact of a range of design and technology activities on the individual, society, and the environment through the development of projects.	■ environmental and social issues including – personal values – cultural beliefs – sustainability – safety and health – community needs – individual needs – equity.	■ assess the impact of the activities undertaken in the development of design projects on the individual, society and the environment. ■ evaluate examples of design and production and relate these to environmental and social issues.

5.1 Environmental issues

Environmental issues include:

- the greenhouse effect or global warming
- the ozone layer
- tropical deforestation
- waste
- resource consumption and sustainability
- water pollution
- noise pollution
- cradle to grave to cradle.

The greenhouse effect and global warming

Gases in the atmosphere insulate the earth, preventing the sun's heat, reflected from the earth's surface, escaping into space. The main causes of global warming are the increase of these gases in the atmosphere from industrialisation and agricultural development and the burning of fossil fuels such as wood, coal and oil, which produces carbon dioxide.

The increased production of chlorofluorocarbon (CFCs), nitrous oxide, hydrochlorofluorocarbons (HCFCs) and carbon dioxide from aerosols and motor vehicles has led to an increase in global temperatures, climatic changes, rising sea levels and the redistribution of land suitable for agricultural production.

Design considerations

The following are design considerations and possible solutions to the issue of the greenhouse effect and global warming.

- Design a product, system or environment that absorbs the gases that are contributing to global warming. Forests absorb carbon dioxide from the air and this environment is currently being simulated in human-created ecosystems that are designed to help solve the planet's climatic issues.
- Design energy-efficient products that educate and inform the consumer about the greenhouse effect. Ensure that these are economically accessible to the consumer.
- Design products for recyclability rather than planned obsolescence. The products may have replaceable parts or use new materials and/or emerging technologies.
- Design alternative power sources that do not rely on the use of fossil fuels. This may involve the use of solar power as an alternative energy source.
- Improve energy conservation and efficiency in redesigning machinery and production techniques in industry.
- Use energy-efficient materials to reduce power usage in homes. Insulation and solar power in a house will reduce the use of fossil fuels.

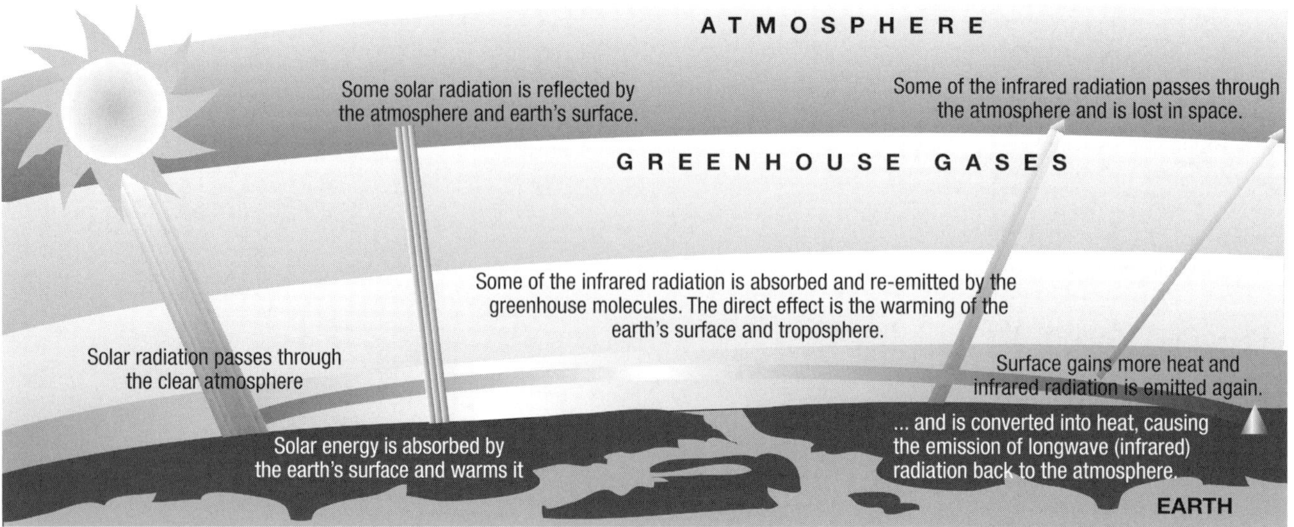

Figure 5.1 Greenhouse effect

- Reduce dependency on cars in order to reduce gas emissions. There is much research and development in this area, with hybrid cars that run on both fuel and electricity being very popular. In Beverly Hills, California, it is possible to park your car and plug it in to recharge it.

The ozone layer

Ozone gas occurs naturally in the stratosphere at 12–50 kilometres above the earth's surface. It forms a protective shield from the ultraviolet (UV) radiation of the sun. If the ozone layer thins or breaks more UV radiation reaches the earth's surface, causing damage to living organisms. Increased UV rays cause changes to the climate and to ecosystems.

Ozone is broken down naturally by UV rays but this is accelerated by the presence of chlorine which is found in CFCs. CFCs are produced in the manufacturing of blowing foams used in insulation and packaging materials, in aerosol sprays, and in solvents for cleaning electronic components. These gases cause this layer to break down, allowing dangerous UV rays to enter the atmosphere. Currently there is a hole in the ozone layer over the South Pole which is the size of Australia and is contributing to the melting of the polar ice caps.

Design considerations

The following are design considerations and possible solutions to the issue of the ozone layer.

- Use alternatives to CFCs and HCFCs. A number of more environmentally friendly gases are available but the cost is a little higher.
- Alternative packaging materials are available and companies such as McDonalds have completely changed their packaging to an environmentally friendly approach.
- Alternatives to aerosol products such as roll-ons are available. The environmental advantages of these must be further promoted to consumers to increase their popularity.
- Alternative insulation materials are being designed. These do not use the carbon-based gases that are destroying the ozone layer. Some of these materials are being manufactured from recycled paper.
- Alternatives for refrigeration and air-conditioning gases need to be considered. The rating scales on these products inform consumers of their environmental impact but the question to be considered is should there be an environmental tax for products that are not environmentally friendly? This would increase the cost, making them unappealing to consumers and inappropriate to designers.

Tropical deforestation

The main cause for concern is the rate of destruction of the tropical rainforests. If the current rate continues it is estimated that all rainforests will be destroyed in the next sixty years.

The effects of deforestation include the extinction of species and the destruction of ecosystems; the disruption of local climates, possibly leading to desertification; increased temperatures and global climate changes due to changes in rainfall patterns; and the loss of habitat for local people. The destruction is also a significant contributing factor to the greenhouse effect and brings about global temperature change because forests absorb carbon dioxide.

Figure 5.2 Amazon deforestation

Tropical deforestation is caused by a variety of factors including population growth, which is causing people to cultivate forest areas, the fuel required to support the population's energy requirements, commercial logging used to generate foreign currency and consumer demand for exotic timber products.

Figure 5.3 Australian deforestation

Design considerations

The following are design considerations and possible solutions to the issue of tropical deforestation.

- The use of sustainable forests whereby forests are now being planted to replace those that have been felled in order to meet consumer demand.
- The purchase of areas of tropical forests by individuals and companies in order to maintain these areas for future generations.
- Consumer education about the origin of materials and processes used in the manufacture of furnishings. This could be in the form of labelling laws.
- The use of timber as a global 'currency' in developing countries could be more fully considered by designers and alternative raw materials sought.

Waste

Developed countries produce over one billion tonnes of waste each year, with the average Australian household producing one tonne each. Most of this ends up as landfill, is incinerated, or is dumped on the earth or at sea.

The space for landfill sites is running out and waste is being transported over long distances, making the disposal both economically and environmentally costly. Rubbish does not biodegrade into harmless substances which assimilate into the soil and the materials often contain contaminants that leach into rivers and thereby into our drinking water. In addition, the gases released contribute to global warming unless they are tapped and used for heating purposes.

Burning waste generates energy in the form of heat. To use this energy one must be close to the source of the fire or the heat is lost and dispersed into the ozone layer. The burning of waste causes pollution and creates toxic gases unless it is burnt at very high temperatures. Plastics and chemicals such as pesticides give off dioxins which are extremely toxic. The residue that is left after incineration can contain dangerous metal pollutants which must then be collected and disposed of but these are often buried.

Depositing rubbish in the earth and dumping it at sea damages both the soil and marine life. There is a limit as to how much can be dumped or buried and a real danger of poisonous contaminants leaking from damaged containers.

Figure 5.4 Industrial pollution

Figure 5.5 Waste pollution

Design considerations

The following are design considerations and possible solutions to the issue of waste disposal. Currently the most effective way to resolve this problem is to produce less waste. This can be done in a number of ways including:

- increase the product life so that it lasts longer and does not need to be replaced as often.
- reduce the amount of material used in packaging.
- use biodegradable materials in the product and packaging. Constantly re-use 'green' shopping bags made from non-biodegradable nylon.
- reuse, recycle and remanufacture as often as possible.
- consider materials, components, planned obsolescence and replaceable components when designing a product, and design exterior elements that do not date.
- design a new waste removal system. This has been the life work of a Canadian man, Mr John Thompson, who purchased landfill sites and converted the landfill to power plants that supplied electricity to local towns. He has now designed a product the size of a large air-conditioner that is attached to a home and which converts household garbage into electricity. He is facing huge opposition to the release of this product as governments realise that their revenue from electricity bills may be reduced.

Resource consumption

The conservation of natural resources and the responsible management of renewable resources are at the centre of sustainable development. There is a need to aim for meeting today's needs without harming the future generation's ability to meet their needs. In doing this we are minimising our own carbon footprint.

Design considerations

The following are design considerations and possible solutions to the issue of resource consumption. Designers can impact on this in a number of ways as in the following.

- Materials may be natural, synthetic, recycled, virgin, renewable or non-renewable. The decision which to use must be thoroughly considered to select materials with the lightest carbon footprint. There are difficult trade-offs when making this decision but it is important to select raw materials close to where they will be used to save transportation costs and the use of fossil fuels. Materials from non-renewable sources should be recycled. The extraction process of the raw materials must also be considered in terms of the impact on the local environment.
- Consideration must be given to the type and amount of energy used in extraction, manufacture, transit, use and disposal.
- A reduction in the need to consume could be achieved by creating minimalist products or multipurpose items.

Water pollution

The growth in population and increased use of water for industrial purposes means that there is an insufficient supply of clean water to meet demand. In developing countries water is polluted from sewerage, nitrates and chemical cocktails seeping from landfill sites and industrial discharges. This pollution is a problem for humans, plants and animals.

Pressure exists in industry for a 'closed loop' waste management system where no harmful chemicals can leach into water tables, but large volumes of water are required for this so it may not be practical.

The use of chlorine to bleach paper has been criticised and paper mills are now using less toxic hydrogen peroxide. The use of dyes in textile mills produces harmful emissions which are not fully biodegradable.

Design considerations

The following are design considerations and possible solutions to the issue of resource consumption.

- The suppliers of manufactured products should be inspected to ensure that they are not unnecessarily polluting.
- The industrial processes should be checked and redesigned to meet environmental standards.
- Manufacturing materials need to be 'green' and these could be made cheaper by taxes being placed on materials that do not meet environmental standards.

- Saving water should be as important as saving energy to industries, communities and individuals.
- Household appliances could be redesigned as water-saving devices. Some of these already exist, such as water-saving shower heads and incorporating systems that allow households to recycle the grey water output from washing machines and use it to water gardens.

Noise pollution

Although noise may not be life threatening it causes discomfort to some people. Much improvement has been made in reducing the noise in industry, manufacturing and machinery. Domestic machines such as dishwashers and lawn mowers are being produced with reduced noise levels. Air traffic is still a serious problem for people living under aircraft flight patterns, although increased insulation can assist with this issue.

Cradle to grave to cradle

The term 'life cycle' refers to the balanced and holistic evaluation of a product. This requires the assessment of raw materials, and the product's manufacture, distribution, use and disposal, including all transportation factors. The total of all these steps is the product's life cycle.

A life cycle assessment (LCA) is also known as a life cycle analysis, eco-balance or cradle to grave analysis. It is the investigation and valuation of the environmental impacts caused by the existence of a product or service.

The goal of LCA is to compare the full range of environmental and social damages assignable to products and services and thereby choose the least burdensome one. At present it is a way to account for the effects of technologies responsible for the production of goods and services.

The life cycle or cradle to grave impacts include the extraction of raw materials; the processing, manufacturing and fabrication of the product; the transportation or distribution of the product to the consumer; the use of the product by the consumer; and the disposal or recovery of the product after its useful life.

The cradle to cradle approach demands that designers take the process one step further, ensuring that the products are either renewable, reusable or recycled. This cradle to cradle system underlies all assessment for eco-labelling systems. Cradle to cradle is the new goal for most manufacturers.

There are four linked components of LCA:

- Goal definition and scoping—identifying the LCA's purpose and the expected outcomes of the study, and determining what is and is not to be included.
- Life cycle inventory—measuring the energy and raw material inputs and environmental releases associated with each stage of production.

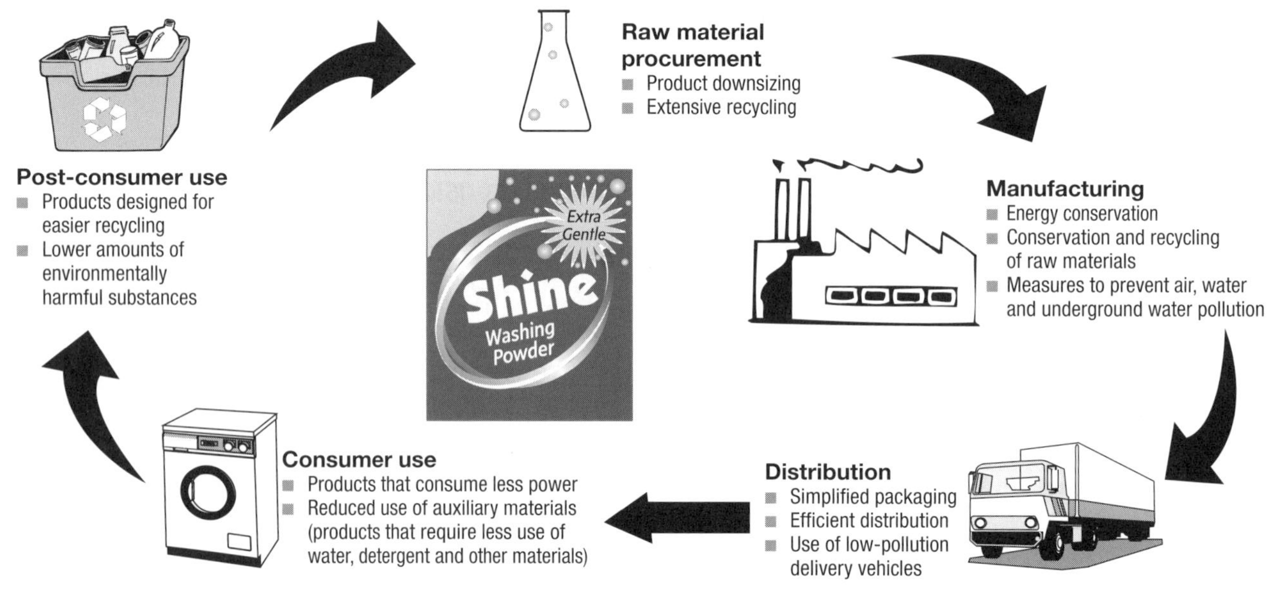

Figure 5.6 Life cycle analysis (LCA)

- Impact analysis—assessing the impacts on human health and the environment associated with energy and raw material inputs and environmental releases.
- Improvement analysis—evaluating opportunities to reduce energy, material inputs or environmental impacts at each stage of the product's life cycle.

LCAs are vital to support the development of eco-labelling schemes which are operating or planned in a number of countries. In order for eco-labels to be granted to chosen products the awarding authority needs to be able to evaluate the manufacturing processes involved, the energy consumption in manufacture and use, and the amount and type of waste generated.

Whichever name is used to describe it, LCA is a potentially powerful tool which can assist regulators to formulate environmental legislation, help manufacturers analyse their processes and improve their products, and perhaps enable consumers to make more informed choices.

What is a life cycle analysis?

Taking the case of a manufactured product, an LCA involves making detailed measurements during the manufacture of the product, from the mining of the raw materials used in its production and distribution through to its use, possible reuse or recycling, and its eventual disposal.

LCAs enable a manufacturer to quantify how much energy and raw materials are used, and how much solid, liquid and gaseous waste is generated at each stage of the product's life. This study would ignore second-generation impacts, such as the energy required to fire the bricks used to build the kilns used to manufacture the raw material.

One cannot prove via LCAs that one product or one process is better than any other since many parameters cannot be simplified enough to reach this conclusion.

In the case of manufactured goods, the most important time for LCA information to be taken into consideration is at the design stage of new products.

Where LCA is used to evaluate procedures rather than products the information can help ensure appropriate choices are made.

5.2 Social issues

5.2.1 Personal values

Values may be defined as something that is important to someone. They may be tangible or a part of a personal belief system such as integrity, creativity and honesty.

When a client selects a designer they usually do so after ensuring that the designer holds the same personal values that they do. A client will look at the range of a designer's work before deciding which one to use. These values may be personal, functional or aesthetic.

Personal values may relate to the status or reputation of the designer in the community, their approach to 'green' design, or their honesty, integrity and creativity. The functional values may relate to the designer's previous experience, the quality of their work, the time taken to complete the project and the cost of the project. Aesthetic values refers to the appearance of the final product.

A designer may be selected because of the materials, tools, techniques and design styles or ideas that they use. A designer may value the use of natural timbers in traditional styles. If the client aligns with the designer's values then it is likely that the designer will be selected.

5.2.2 Cultural beliefs

Cultural beliefs are common understandings, traditions, mores and folkways held by a group of people from a similar ethnicity. These may be 'truths', misconceptions or superstitions. When designing a product, system or environment it is essential to take into account the client's cultural beliefs in order to ensure that the design is a success.

An example of this would be the MGM Grand Hotel and Casino in Las Vegas. The MGM Grand had an amazing green lion's head with a wide mouth which was used as the entrance to the hotel but they had difficulty attracting Chinese customers. A Chinese man advised them that the reason for their lack of customers was because it is considered bad luck for a Chinese person to walk through the mouth of a lion. The Chinese customers avoided this casino because they believed that if they gambled there they would lose. The hotel owners considered this belief before hiring architects to redesign the entrance to the hotel and decorating it with a red and gold interior. These colours were chosen because the Chinese believe red to be a lucky colour and the gold symbolised prosperity. This change in both interior and exterior design was aimed at meeting the Chinese customers' cultural beliefs in order to encourage them to enter the hotel and casino. The changes had the desired effect and the customers returned.

5.2.3 Sustainability

The definition of sustainable design is 'delivering the best performance for the least cost'. This has not always occurred throughout history. Amory Lovins (2004) reminds us that the Industrial Revolution emphasised labour productivity while nature was abundant, whereas the next industrial revolution will unfold on the basis of people being abundant and nature and food scarce which is why redesign must occur urgently if our planet is to sustain life as we know it.

Consider that nearly every creature except humans builds unobtrusive or hidden 'homes'; every creature except humans has solar energy as its sole energy source; every creature except humans recycles all its waste, and not just some of it; and every creature except humans uses sustainable technology to gather its food and to manufacture products such as shelter and systems for gathering and storing food items.

The world needs to become a sustainable place because the United Nations predicts that there will be nine billion people by 2050 and therefore more resources will be diverted into consumption, leaving less for productivity, and already limited resources will be stretched to the limit with demands on health, education and nutrition growing.

Conventional design and construction methods have been linked to environmental damage including depletion of natural resources, air and water pollution, toxic wastes and global warming. It is time to make this a sustainable planet.

5.2.4 Safety and health

For information on this topic refer to Chapters 2, 8 and 16 of this study guide.

5.2.5 Community needs

Community needs may also be physical, sociological or emotional and these needs drive design projects. The design projects are created for either a member of the community or with other community members.

Physical community needs may involve the desire to create sporting grounds, clubhouses, parklands or hostels, or to manage clothing, food and toy collections to assist those less fortunate.

Sociological community needs may involve fund-raising activities such as helping individuals or the community as a whole after a natural disaster or for helping community members with health and other problems.

Psychological community needs may involve collecting books to be sent to a developing country or to build or repair community buildings such as schools and hospitals.

5.2.6 Individual needs

A 'need' may be defined as something that we must have in order to survive. It is something that we cannot do without. A need is not to be confused with a 'want'. A want is something that we would like to have but is not essential for survival. Individual needs may be personal or common to a group of people. They may be physical, sociological or psychological.

Physical needs include those for food, air, water, shelter, protection and sleep. Food is needed to nourish the body, air to breathe, water to drink, shelter from the weather, protection from harm and sleep to renew the body.

Sociological needs are those that relate to our emotions and feelings about various issues. Sociological needs include the need to belong, to feel safe and for a confident self esteem. It is important to belong to a family, kinship or peer group to encourage a sense of belonging. Belonging is a part of feeling safe and having support or emotional security. Self esteem is needed to allow interaction between people.

Psychological needs are those that relate to our own development. They include the needs for intellectual stimulation, for a positive self concept, the need to achieve, and to believe in something or someone, be it a spiritual being or simply yourself.

Individual needs must be explored when considering a design situation. The need must be real and be based in a problem. Examples of problems stemming from individual needs that you could use as design projects include:

- a blind person has difficulty matching coloured tops and bottoms in their clothing
- an arthritis sufferer may have difficulty using a knife, fork, scissors, and so on
- an older person may not be able to see the numbers on a telephone
- a person who has had a knee operation may not be able to climb stairs
- a tourist may not be able to read street signs in a country where they do not speak the language
- a farmer may find it time consuming to travel kilometres to the mail box each day
- a teenager may have five guitars and nowhere to store them so that they are easily accessible.

5.2.7 Equity

The Australian Environment Protection Authority (EPA) has developed an environmental equity policy that states no segment of the population should, because of its racial or economic make-up, bear a disproportionate share of the risks and consequences of environmental pollution nor be denied equal access to environmental benefits.

This ensures equal protection from environmental hazards for individuals, groups or communities regardless of race, ethnicity or economic status. It promotes the ideal of equal treatment and protection for various racial, ethnic, and income groups under environmental statutes, regulations, and practices applied in a manner that yields no substantial differential impacts relative to the dominant group.

Environmental justice is the fair treatment and meaningful involvement of all people regardless of race, color, national origin or income with respect to the development, implementation and enforcement of environmental laws, regulations and policies. The EPA has this goal for all communities and persons across Australia. It will be achieved when everyone enjoys the same degree of protection from environmental and health hazards and equal access to the decision-making process in order to have a healthy environment in which to live, learn and work.

Key concepts and definitions

Carbon footprint—a carbon footprint is the total set of greenhouse gas emissions caused by an organisation, event or product. For simplicity of reporting it is often expressed in terms of the amount of carbon dioxide or its equivalent of other greenhouse gases emitted.

Global warming—this is the effect of an increase in global temperatures caused by the greenhouse effect and ozone depletion. Related events include melting polar ice caps, unpredictable and extreme weather patterns, the extinction of plant and animal species, rising sea levels and the redistribution of land suitable for agricultural production.

Greenhouse effect—this is where gases in the atmosphere that are caused by industrial and agricultural development and the increased production of chlorofluorocarbons and the burning of fossil fuels insulate the earth and prevent the sun's heat from escaping, thereby causing a 'greenhouse' effect.

Life cycle—a fair, holistic evaluation of a product. This requires the assessment of raw materials, and the product's manufacture, distribution, use and disposal, including all transportation factors.

Ozone layer—ozone gas occurs naturally in the stratosphere at 12–50 kilometres above the earth's surface. It forms a protective shield from the ultraviolet (UV) radiation of the sun. If the ozone layer thins or breaks more UV radiation reaches the earth's surface, causing damage to living substances. Increased UV rays cause changes to the climate and to ecosystems.

Sustainable design—the conservation of natural resources and the responsible management of renewable resources are at the centre of sustainable development. It is delivering the best performance for the least cost, with the 'cost' being environmental as well as financial.

Tropical deforestation—the cutting down of tropical rainforests. The reason behind this is population growth and consumer demand.
Effects include the destruction of plant and animal species and ecosystems and climatic impacts such as desertification.

Useful websites

- www.designandculture.org/index.php/dc
- www.edcmag.com/
- www.environmentdesignguide.net.au/
- www.socialdesignsite.com

Classroom activities

1. Select one cultural or social group and identify five problems that are specific to them.
2. Investigate one of these cultural or social groups and further investigate one of the associated problems.
3. Working in pairs, brainstorm possible solutions to the problem chosen above.
4. Create a PMI (plus, minus and interesting) for each possible solution.

Key concept questions A p. 201

1. List ten of your personal values and then rank their order of importance.
2. Consider your cultural heritage and list cultural beliefs that are a part of your way of life.
3. Sketch three possible solutions to each of the problems below.
   - A blind person has difficulty matching coloured tops and bottoms in their clothing.
   - An arthritis sufferer may have difficulty using a knife, fork, scissors, and so on.
   - An older person may not be able to see the numbers on a telephone.
   - A person who has had a knee operation may not be able to climb stairs.
   - A tourist may not be able to read street signs in a country where they do not speak the language.
   - A farmer may find it time consuming to travel kilometres to the mail box each day.
   - A teenager may have five guitars and nowhere to store them so that they are easily accessible.

Sample Preliminary questions

Objective-response questions A p. 201

Circle A, B, C or D. (1 mark each)

1. Factors that impact on environmental sustainability include
   A climate change, cyclones, storms and tsunamis.
   B resource depletion, waste disposal, deforestation.
   C renew, reuse and recycle.
   D resource depletion, waste disposal, deforestation, climate change, water and air pollution, ozone depletion and the greenhouse effect.

2. Ozone depletion is
   A the thinning of a hole in the ozone layer.
   B warm gasses trapped beneath the ozone layer.
   C cutting down oxygen-providing trees.
   D the depletion of oxygen due to gases trapped near the earth.

Short-answer questions A p. 201

3. Differentiate between cradle to cradle and cradle to grave. (2 marks)
4. Why is deforestation an issue? (3 marks)
5. What does the debate surrounding climate change centre on? (4 marks)

Extended-response questions A pp. 201–202

6. Consider a design project that you have completed and evaluate its environmental and social impact on society by answering the following questions.
   (a) State the design situation and the problem to be solved. (3 marks)
   (b) List the values held by the client that impacted on the final design solution. (2 marks)
   (c) Determine if it is an individual or community-based problem and how you know which it is. (4 marks)
   (d) Explain how equity was ensured in this project. (6 marks)
7. Complete the following environmental report.
   (a) Describe the problem and your produced solution. (1 mark)
   (b) Explain how the problem contributed to the greenhouse effect or global warming. (2 marks)
   (c) Evaluate the degree that the project contributed to the depletion of the ozone layer. (1 mark)
   (d) If deforestation was an issue suggest how this could have been avoided. (1 mark)
   (e) Evaluate how material and processing wastes were disposed of. (2 marks)
   (f) Assess whether the design solution could be considered sustainable. (2 marks)
   (g) Complete a cradle to grave flow chart of the solution. (2 marks)
   (h) Recommend how it may become a cradle to cradle solution. (2 marks)
   (i) Evaluate whether this project would have a light or heavy carbon footprint. (2 marks)

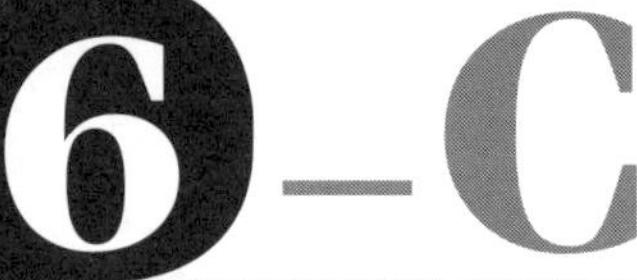

Creative and collaborative approaches in designing and producing

PRELIMINARY OUTCOMES

A student:	You learn about:	You learn to:
P3.1 investigates and experiments with techniques in creative and collaborative approaches in designing and producing.	■ creative approaches including – cognitive organisers – strategies for problem-solving and solution creating – ideas generation. ■ collaborative approaches – design teams: roles and tasks of members – communication between and within design teams – team responsibilities.	■ select and apply a variety of cognitive organisers. ■ apply problem-solving techniques to identified problems. ■ identify the factors that contribute to successful work and collaboration. ■ collaborate and participate in design teams. ■ work cooperatively.

6.1 Creative approaches

6.1.1 Cognitive organisers

Cognitive organisers are ways of making thinking visual! Cognitive organisers are tools that promote higher-level thinking. As you work through your design folio you will need to evaluate and select which tools or organisers to use. Creative problem-solving tools typically consist of software or objects that can be manipulated to facilitate creative techniques.

Active or lateral thinking relates to being involved in problem-based learning, examining problems from all angles, and thinking 'outside the square'. The key to creativity and creative problem-solving is simplicity. In your designs remember to 'keep it simple'.

Students learn to select and apply a variety of cognitive organisers. The following formalised and well-known cognitive organisers combine a range of creative problem-solving techniques.

Mind mapping

Mind mapping is a technique that both reframes the situation and fosters creativity. It is a visual representation of a large number of ideas. It involves drawing the central theme in the centre of the page and branching the main ideas out from it, with other ideas stemming from the central one. A mind map could be used in your design folio to show the range of ideas that you are going to explore, to demonstrate the link between your thoughts on a topic or to show how logical your thinking is. Figure 6.1 is an example of a time management mind map.

Brainstorming

Brainstorming is a group activity designed to increase the quantity of fresh ideas. Getting other people involved can help increase knowledge and understanding of the problem and help to reframe it. In brainstorming sessions a group of people sit around and bounce ideas off one and other, sketching and writing ideas to be followed up at a later date. While brainstorming seldom yields major innovations, many heads can be better than one. You will use brainstorming in your design folio when you list a range of possible design solutions. Types of brainstorming techniques include the following:

- Group chats on the Internet or an internal network allows a group of people to type and view ideas on their computer. During these brainstorming sessions ideas come up fast and furious and are limited only by your typing ability. The unconventional part of this technique is that it allows participants a certain degree of anonymity.

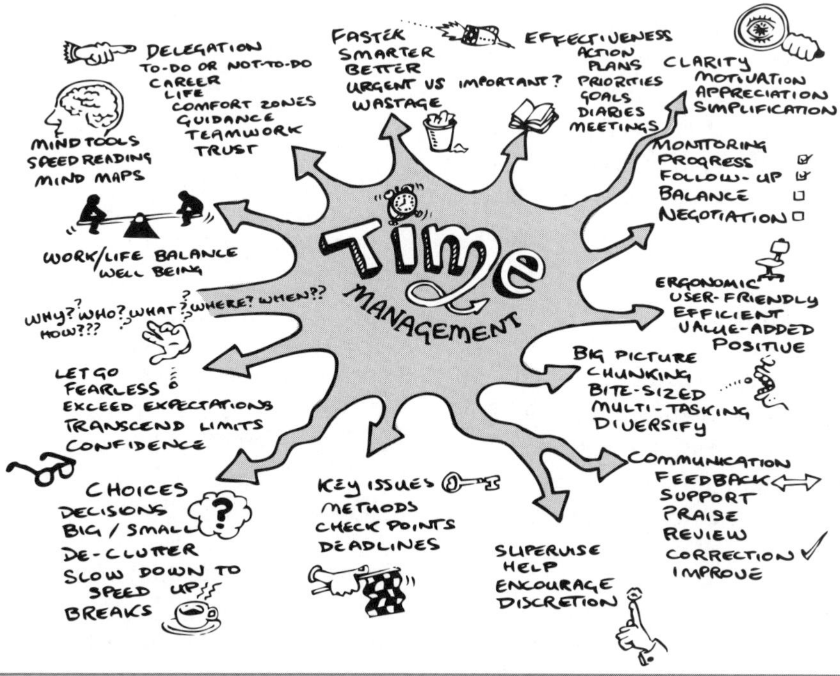

Figure 6.1 Example of a mind map

- The 'pass the idea' technique has each person starting an idea with a sketch or description and then passing it on the next person. Everyone takes turns building on the idea as it gets passed around. By the time you are finished, the ideas have taken as many different directions as there are people in the group.
- In role-playing each person assumes a certain character or personality and acts out the role within a brainstorming session. This technique helps people 'come out of their shells' and reduces inhibitions in general. For example, you could take on the role of a pregnant mother of five who needs to design a baby mover. It is a little like dressing up in a costume and acting and the results can therefore be a little unexpected!
- 'Think, pair, share' is a smaller version of brainstorming. Here people sitting in close proximity to each other think about the situation and note, in point form, their ideas and then share their thoughts with the persons sitting near them.

Prototyping

Another type of cognitive organiser that can be used to demonstrate higher-order thinking skills in your folio is prototyping. Prototyping involves making a simple miniature model, which may or may not be functional, in order to gain a 3-D visual idea of what the project will look like. Prototyping is a way of making progress when challenges seem insurmountable in an especially complex project. One approach is to create a working prototype of a problematic part of the design as this will allow experimentation at minimal cost. Focused prototyping helps resolve small but critical problems one by one. Once you manufacture your prototype you will be surprised at how easily some of the solutions appear.

PMI and the Six Thinking Hats

Edward de Bono is well known for his ability to promote creative problem-solving thinking. He coined the term 'thinking outside the square', also known as tangential or lateral thinking. Some of the cognitive organisers that he created are the PMI and the 'Thinking Hats'.

PMI is a cognitive organiser that is used to evaluate ideas. De Bono defines PMI as Plus, Minus and Interesting points, but in Design and Technology PMI refers to Plus, Minus and Improvements to be made.

PMI may be used in your folio to evaluate a range of existing products. Each product is analysed and the positive and negative points listed along with a list of any interesting points or possible improvements. These qualities may be listed in a table such as the one below.

Design	Plus	Minus	Interesting/ Improvements

Figure 6.2 Example of a PMI table

The list of positives becomes the list of desirable traits for your design solution, the minuses are the undesirable qualities, and the list of interesting points or areas for improvement provide the degree of difference between your design and currently existing design solutions.

The Six Thinking Hats are used to move your thinking forward. De Bono believes that debating or arguing over a topic is a waste of time because no-one considers the other person's ideas and the thinking goes around in a circle. Using the Thinking Hats ensures that everyone is thinking in the same mode at the same time, allowing each person's thinking to move forward.

Figure 6.3 The Six Thinking Hats

You can separate thinking into six distinct categories. Each category is identified with its own coloured metaphorical 'Thinking Hat'. By mentally wearing and switching 'Hats' you can easily focus or redirect thoughts, conversations or meetings.

Advantages of the Thinking Hats system is that they:

- allow for meetings without emotions or egos leading to bad decisions
- avoid easy but mediocre decisions by digging deeper
- increase productivity and lead to more effectiveness
- make creative solutions the norm
- maximise and organise each person's thoughts and ideas
- get to the right solution swiftly and with a shared vision.

Attributes listing

Attributes listing is a great technique for ensuring all possible aspects of a problem have been examined. Attributes listing is breaking the design problem down into smaller bits and seeing what you discover when you do. By thinking of more alternative ideas for the attributes, innovative design is made easy and new and improved products can be inspired.

Feature	Functional characteristic	Attributes	Ideas

Figure 6.4 Example of an attributes listing table

SCAMPER

SCAMPER is a technique that discovers ways to improve attributes. Alex Osborn, a pioneer in facilitating creativity, developed this list of 'idea-spurring questions'. They were later arranged by Bob Eberle as the mnemonic SCAMPER:

- S = Substitute? (other ingredients, materials, and so on)
- C = Combine? (blend, and so on)
- A = Adapt?
- M = Modify, magnify or miniaturise?
- P = Put to other uses?
- E = Eliminate?
- R = Reverse (roles, and so on) or rearrange? (patterns, pace, and so on).

To use SCAMPER you describe the key attributes or components of a situation you wish to change and then apply the SCAMPER questions to generate ideas for achieving your goal.

For example, a bicycle has the following components: frame, pedals, drive sprocket, chain, brakes, tyres, handlebars, and so on. By applying SCAMPER, the following ideas for improving the bicycle were generated:

- much lighter weight frames based on new materials
- pedal grips that strap to secure feet better
- stronger chains with special clamps for easier changing
- improved derailleur gears for rear sprocket
- racing handlebars for a more ergodynamic racing position
- new rear wheel materials to replace spokes.

Use SCAMPER in your folio to generate a range of possible solutions to your client's problem.

The Creative Problem-Solving Process (CPS)

The Creative Problem-Solving Process (CPS) is a six-step method developed by Alex Osborn and Sid Parnes that alternates convergent and divergent thinking phases.

Each phase in CPS has two phases: a divergent phase (D phase) and a convergent phase (C phase). The goal of the D phase is to generate as many ideas as possible. The rules are that there is to be no judgement or evaluation during the D phase so you can and should go wild and wacky. Try to generate as many ideas as possible, even if they are thoroughly impractical: indeed it is better if they are. The goal of the C phase is to carefully and systematically narrow down your options to just one. Your mindset is just the opposite of what you employed during the D phase.

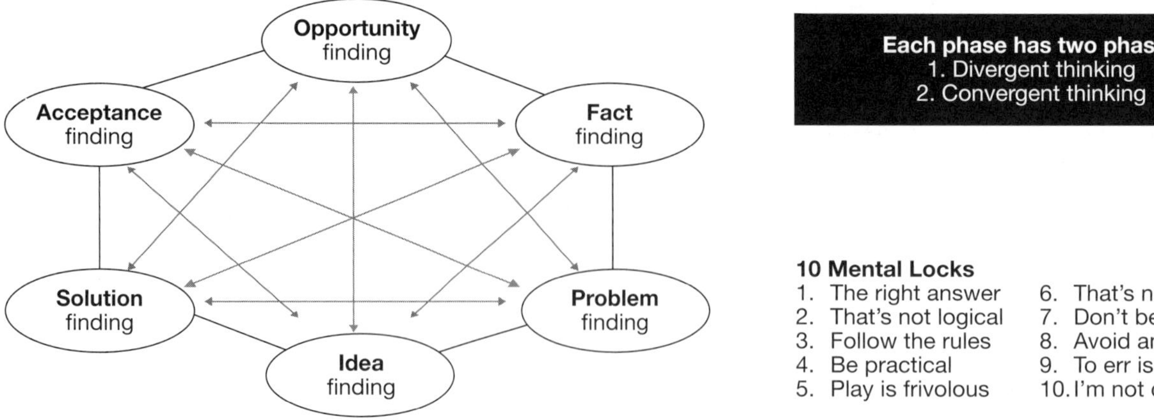

Figure 6.5 The Creative Problem-Solving Process (CPS)

In CPS, formulating the right problem is as important as developing solutions. Well-written problems:

- consider in what ways might I/we benefit from this problem
- are worded in an actionable form because the goal is to move forward towards a solution
- ensure that you have a clear action plan at the end.

Use CPS in your folio to generate a range of possible solutions to the problem and then to narrow down the ideas to a final solution.

Cross-clarification charts and Venn diagrams

Cross-classification charts involve bringing two lots of information together in order to achieve something new. In your design folio this is where two pieces of charted data are overlayed to create new data. This synthesis can produce new, high-level thinking as long as the data on either axis is of a high enough level.

Venn diagrams compare and contrast data in a visual way by showing all overlaps as in the following example.

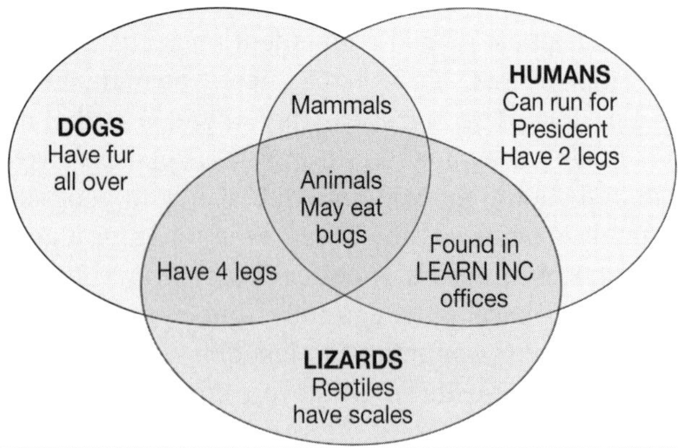

Figure 6.6 Example of a Venn diagram

Thinker's Keys

Thinker's Keys were first introduced by Tony Ryan in the 1980s and are an effective way to introduce different ways of higher-order thinking and promote lateral thinking. Thinker's Keys are included as part of a Bloom's 'Taxonomy of Thinking' and Gardner's 'Multiple Intelligence' approaches to teaching and learning. See Chapter 21 for more information on this topic. Only some keys will be suitable for use in your design folios: you must decide which will be best to use.

A digital version of this resource is available at www.thinkerskeys.com.

Questioning

The importance of questioning cannot be underestimated. Thinking without questioning is like drinking without swallowing. There is plenty of thinking that never achieves lift-off, never contributes to understanding or casts light on issues of importance. That is because thinking can be done in an unquestioning manner. It therefore makes sense to spend time learning how and when to question. High-level questioning can be carried out using Socratic thinking.

The Classical Greek philosopher Socrates believed that if you thought about a problem long enough the answer would come. This does not mean you sit and wait: quite the contrary, if you have a problem to investigate Socratic thinking involves asking everyone you know how they would resolve the issue. You then take the best ideas from each of the people and put them together to come up with the best possible solution.

6.1.2 Strategies for problem-solving and solution creating

Creative problem-solving is the mental process of creating a solution to a problem. It is a special form of problem-solving in which the solution is independently created rather than learned with assistance. Creativity requires innovation as a characteristic. Creativity can involve using new materials in new ways, creating new manufacturing techniques, combining ideas in original formats or synthesising completely new ideas. Creativity equals innovation, not imitation.

The solution must either have value, clearly solve the stated problem, or be appreciated by someone for whom the situation improves. The situation prior to the solution does not need to be labelled as a problem. Alternate labels include a challenge, an opportunity or a situation where there is room for improvement.

Innovations begin as creative solutions, but not all creative solutions become innovations. Some innovations also qualify as inventions. Inventing is a special kind of creative problem-solving in which the created solution qualifies as an invention because it is a useful new object, substance, process, software or other kind of marketable entity. Creative problem-solving strategies can be categorised as below.

Creativity techniques are designed to shift a person's mental state into one that fosters creativity. These techniques are described as cognitive organisers. One technique is to take a break and relax or sleep after intense thinking about a solution.

Cognitive organisers are designed to reframe the problem. For example, reconsidering one's goals by asking 'What am I really trying to accomplish?' can lead to useful insights. They are designed to increase the quantity and quality of fresh ideas. This approach is based on the belief that a larger number of ideas increases the chance that one of them has value.

Creative problem-solving techniques are designed to foster a fresh perspective that makes a solution obvious.

6.1.3 Ideas generation

An integrated design process is proving to be the key to achieving high-performance projects. By bringing all of the project stakeholders to the table at the early conceptual stage the design team can proceed with a common set of goals.

6.2 Collaborative approaches

The following factors influence successful work and collaboration. Factors related to the context are:

- the history of the collaboration
- the collaborative group must be seen as a leader in the community
- favourable political and social climate.

Factors related to membership characteristics are:

- mutual respect, understanding and trust
- appropriate cross-section of community members
- members see collaboration as in their self-interest
- ability to communicate
- ability to compromise.

Factors related to process and structure are:

- members share a stake in both the process and the outcome
- multiple layers of participation
- flexibility
- development of clear roles and guidelines
- adaptability
- appropriate pace of development.

Factors relating to communication are:

- open and frequent communication
- established informal relationships and communication links.

Factors relating to purpose are:

- concrete attainable goals and objectives
- shared vision
- unique purpose.

Factors relating to resources are:

- sufficient funds, staff, materials and time
- skilled leadership.

6.2.1 Design teams: roles and tasks of members

A functional design team consists of the following:

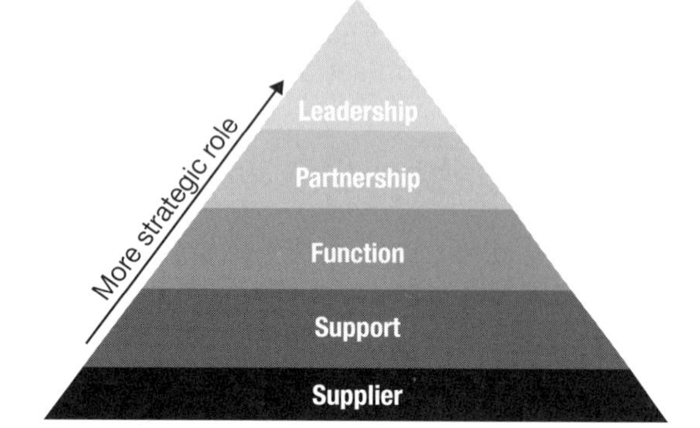

Figure 6.7 The design functionality axis

Designers often complain that their work is not recognised or respected in their organisation. What many designers do not realise is how their designs are treated is related to how design is viewed within their organisation. In design management methodology most organisations have in-house design teams but also use consultancies.

There are many levels of how design is integrated within an organisation. In 'the design functionality axis' the axis stretches from a low level supplier relationship at the base of the triangle to a strategic design leadership relationship. The higher up the pyramid you are positioned, the greater function you have.

6.2.2 Communication between and within design teams

The client plays an important role in reviewing and approving design proposals and revisions. The materials selection process is equally important. In the first few meetings the client's needs, budget, taste, lifestyle and goals will be discussed in depth. The information garnered from these conversations becomes the source of inspiration from which a designer will prepare preliminary design sketches.

To help better communicate thoughts and goals to a designer prepare a journal of sketches, images and ideas. Create lists to record your thoughts: 'new' will list conditions and elements that are ideas and 'old' will list conditions and elements that already exist and you wish to retain in the design solution. Then, prioritise the items in both lists by order of importance to determine which aspects you will include first. Clients are important members of the design team too and the combined journal and comments will help define the design task.

6.2.3 Team responsibilities

A supplier relationship

This is the lowest relationship denominator and is at the bottom of the pyramid. In a supplier relationship, suppliers are often put in an 'execution role' of supplying to the designers anything they require, such as materials, prototypes, research, and so on. In other words, a design is treated a means to an end and is often seen as being done cheaply and quickly. The majority of consulting and freelance work often operates at this level. With in-house designers, it is rarely different. Here a designer mostly works alone, does almost everything remotely creative, and has a hard time trying to convince his or her boss the value of design.

A supporting role

Designers are seen as very tactical in nature. In other words, design is seen as useful in articulating a required need. Designers are found as part of, and doing the bidding of, established departments such as research and development, engineering, manufacturing and even marketing. While in a better position than a supplier, a designer in a supporting role still suffers from working with second-hand information, and has very little influence in the brief requirements. Designers have to deal with the departmental politics of diverging objectives, a difficult design decision-making process, and the need to rework designs due to changing requirements.

In a functional position

Strong designers with a great track record, or who have been in a company for a long time, will often evolve their contribution into a functional role. Here a young organisation starts to mature in regard to their view of design. Such functional roles are evident with the creation or existence of a design department now on par with the other functional departments. Design in a functional role starts to really provide value to a company by having an equal say in solutions. Even though they are closer to the decision maker, designers still need to negotiate different environments to get things done and have little say in decision making.

A partnership

This is probably one of the best places a designer can be in. In a partnership arrangement a designer gets to be part of the decision-making process, as well as influence the business strategy of the company. With close partnerships, a designer commands a lot of trust within an organisation and their word carries a lot of weight. It takes a skilled designer to be successful in this role, as he or she has to frame complex design problems into a language the business can understand.

Strong design leadership

Here the role of design is strategic and design is the main driver in initiatives. Design does this by identifying potential opportunities and articulating solutions that are vital for a successful brand or business. Design must reside in the soul or the 'DNA' of a company as it is integral to its daily operations.

Key concepts and definitions

Cognitive organisers are used to show your thinking. They comprise graphic organisers such as tables and charts.

Collaborative approach—a group of people working together.

Design teams—a small group of people, with specific roles, working together to find a solution that meets the clients' needs.

Ideas generation—processes used to generate thinking and initiatives.

Problem-**solving and solution creating** are used to describe the design process where the designer starts with a client's need and works through the process before arriving at a final design.

Useful websites

- vels.vcaa.vic.edu.au/support/usingcogorg.html
- www.boardofstudies.nsw.edu.au
- www.curriculumsupport.education.nsw.gov.au/schoollibraries/assets/docs/cogorgbklet.doc
- www.edwarddebono.com/Default.php
- www.rinkworks.com/brainfood/p/latreal1.shtml
- www.sunlink.ucf.edu

Classroom activities

1. Convert Chapter 3 on the elements and principles of design and their psychological appeal to a mind map.
2. Brainstorm with your peers ways in which ideas can be generated.
3. Use a think, pair, share technique to problem-solve the following design brief:

Umbrellas are heavy to carry, awkward when getting into and out of automobiles, break easily and are not perceived as 'cool'. Design a device that will protect a young mother and her baby from the rain when stacking groceries into the car.

Key concept question A p. 202

Using a cognitive organiser, show how, in the next six weeks, you plan to resolve any major issues related to your MDP.

Sample Preliminary questions

Short-answer questions A p. 202

1. What is the purpose of cognitive organisers? (2 marks)
2. Discuss the role of creative problem-solving in the design process. (5 marks)

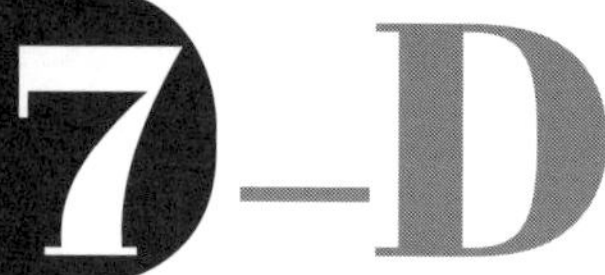

7–Design processes and solutions

PRELIMINARY OUTCOMES

A student:	You learn about:	You learn to:
P4.1 uses design processes in the development and production of design solutions to meet identified needs and opportunities.	■ project analysis – design briefs – the appropriateness of design solutions – the criteria for evaluation and factors to consider. ■ marketing and market research – the purpose of market research – the sources of data and information gathering techniques – the marketing environment.	■ formulate and analyse design briefs. ■ identify the parameters of design. ■ identify criteria for success of design projects. ■ produce functionally and aesthetically appropriate design projects. ■ conduct market research to identify needs and opportunities.
P5.3 uses a variety of research methods to inform the development and modification of design ideas.	■ research methods – qualitative and quantitative research – questionnaires – surveys – interviews – observation – tests and experiments – statistical analysis – information research including print and electronic sources. ■ interpreting and presenting data. ■ ethics in research.	■ select and use a variety of research methods to inform the generation, modification, and development of design ideas. ■ analyse, interpret and apply research data to the development of design projects.

7.1 Project analysis

7.1.1 Design briefs

The design brief is the most important ingredient when starting a successful major design project. If the design brief is well constructed it can shorten the time of the design process; it will also make the process of arriving at an appropriate design solution so much easier. The purpose of the design brief is to clarify the need by developing an understanding of what is to be accomplished. It will continually provide direction throughout the design process and will serve as a benchmark when testing and evaluating through the course of the MDP. By defining goals and objectives the design brief will be a benefit in identifying a course of action while clearly communicating to others your intentions: it will allow you to focus on exactly what you want to achieve before any work starts on the project.

When developing your own design brief you must start with a real problem. A designer may develop their own design brief or may be given one by the client. The design brief is the starting point of a productive cycle of discussions on how the project will evolve.

In developing a design brief you need to clarify the need or problem and define the situational limitations in terms of time, cost, materials and environmental expectations. It is a statement of what a 'not-yet-designed product' is intended to do. The design specifications that need to be met are recognised as design parameters which involve limitations and act as initial boundaries in the development of the design. They will ensure that the subsequent design and development of a product meets the needs of the user. For designers there is a continual management of design parameters. Each designer recognises these design parameters and classifies them as either non-negotiable constraints or negotiable considerations.

Constraints and considerations

To a designer, a constraint involves aspects of the design that are fixed and will not be changed. Here the designer must work with constraints involving budget, size, time, installation restrictions or environmental application of the design.

A consideration will involve aspects that require further thought or research. They may be aspects that are quite obvious from the brief or may surface during the design process where issues related to the type of product being designed or information about the client's needs and tastes and style may become better refined.

Many constraints and considerations will be determined by the design brief and through consultation with the client. However, there are factors that should be considered in all products such as function, aesthetics, ergonomics, safety and ethics. Poor designs often occur as a result of mismanaged constraints. It is necessary that designers recognise each constraint and balance them as part of strong design thinking to determine which are negotiable.

7.1.2 The appropriateness of design solutions

The importance of identifying specific design parameters in the design brief is a critical part of the design process. It will ensure that important design issues are considered and questioned before the designer starts work. It helps develop trust and understanding between the client and designer and serves as an essential process to effective design, production and evaluation. For additional information on this topic refer to Chapters 9, 11, 12 and 20.

7.1.3 The criteria for evaluation and factors to consider

Before the design idea has reached maturity the designer must undergo a process of evaluation, using a list of successful measures or design factors. By forming a set of standards or criteria the designer can judge the merit, worth and significance of the appropriate design. This will validate the idea and provide criteria that will be used at the end to evaluate the design's success.

An evaluation plan outlines the conditions for success and may involve a number of criteria depending on the design.

Criteria to establish success

To develop a set of criteria you need to predict your final design. You must determine what needs to be done and move toward a final solution. To develop the criteria, list all the functional and aesthetic requirements. Determine how these requirements work together to complement the others. Questions may include the following:

- Does the design meet the need of the user or client?
- Has the idea been done before?
- Is the design ethical?
- Is the solution meeting all requirements of functionality, safety and reliability?
- Is the solution aesthetically appropriate?
- Is it at a level of quality that can be discerned?
- Are the materials, tools and techniques used appropriate to the solution?
- Is the solution cost effective and does it fall within time constraints?
- Will the solution make the most of current technology?

7.2 Marketing and market research

7.2.1 The purpose of market research

Market research is a systematic method of collecting and analysing a range of data about a particular target market, competition or environment. The purpose of any market research project is to achieve an increased understanding of the subject matter. With markets throughout the world becoming increasingly competitive companies have realised how research can provide them with the ability to explain, predict, monitor, discover and hypothesise the market. To conduct market research organisations may decide to use their own marketing department or commission it through an agency or consultancy. The most important thing to consider when undertaking any research project is what is expected to be achieved from the research.

While the methodologies of data vary they can be divided into two research categories: primary and secondary research. Both are useful as they differ in their methodologies.

Types of research

Primary research involves the use of first-hand data and includes surveys, interviews and the use of focus groups. Secondary research uses data already collected and analysed by others, for example, articles in journals.

Quantitative research deals with numbers and often involves statistical analysis. Qualitative research includes anecdotal evidence, determined by the respondent. Design students will collect data in much the same way and evaluate, make conclusions and recommendations from their findings, and relate the information to their initial research objectives. For further information on types of research refer to pages 75–77 of this guide.

7.2.2 The sources of data and information gathering techniques

Market segmentation

The purpose of segmenting a market is to allow the marketing/sales program to focus on the 'most likely' target market or demographic. Market segmentation makes it possible for firms to tailor the marketing mix for specific target markets, thus better satisfying customer needs. The identified market segments are summarised by profiles, each segment can be evaluated, and a target market segment selected. A market segment should be:

- measurable
- accessible by communication and distribution channels
- different in its response to a marketing mix
- durable (not changing too quickly)
- substantial enough to be profitable.

A market can be segmented based on variables among groups. Forms of segmentation include:

- geographic segmentation based on region, climate, population density and population growth rate
- demographic segmentation based on age, gender, ethnicity, education, occupation, income and family status
- psychographic segmentation based on values, attitudes and lifestyle
- behavioural segmentation based on usage rates and patterns, price sensitivity, brand loyalty and benefits sought.

Industrial markets are segmented somewhat differently from consumer markets, as described below.

Business market segmentation

While many of the consumer market segmentation bases can be applied to businesses and organisations, the different nature of business markets often leads to segmentation on the following lines.

- Geographic segmentation is based on regional variables such as customer concentration, regional industrial growth rate and international macro-economic factors.
- Customer type is based on factors such as the size of the organisation, its industry, position in the value chain, and so on.
- Buyer behaviour is based on factors such as loyalty to suppliers, usage patterns and order sizes.

Choosing your data-gathering methods

There is a variety of data-collection methods that you may use in developing decision-making information. Particular methods may be chosen for particular use, or all could be used within the collection process. Selection of any data-collection method is based on the merits it will provide you with your evaluation and consequent planning.

Surveys and questionnaires are generally regarded as the same. Surveys and questionnaires are standardised to ensure reliability and are structured in a way to try and remove any form of bias. The style of questioning can vary from open-ended to closed-ended questions. An open-ended question could ask the respondent to formulate their own answer, whereas a closed-ended question has the respondent picking an answer from a given number of options.

As a primary research method, an open-ended interview allows the interviewer to ask the respondent for additional information to obtain details and new insights. There is an opportunity to raise the important issues and to obtain recommendations. The selection of the respondents is vitally important, as are the questions.

An observation is a qualitative method that provides descriptive information about what happens in an activity, process or discussion. It provides an opportunity for an evaluation consultant or other individual outside the organisation to become more familiar with a demonstration of a new product.

Reviews such as company or individual records can be an important part of collected information. Documents routinely collected include work plans, minutes of meetings, grant proposals, annual reports, mission statements, attendance sheets, budget information, correspondence and newsletters. A review of records may provide information about the history and context of the program and about internal and external factors that affect activities. Records may be used to examine changes in a program over time.

Performance testing is designed to measure the knowledge, skills or attitudes that program participants have about a particular topic. Performance tests are often referred to as pre-tests and post-tests.

Community reporting standards, known as 'measures', are documents or reports collected by other groups that contain information and statistics related to the information the designer requires. Community measures may be obtained from government departments at the federal, state and local levels. Accessing this information from departments such as the Australian Bureau of Statistics will help provide data from a larger community context. It may also help in understanding the broader impact of the data beyond your scope.

7.2.3 The marketing environment

The term 'market environment' refers to all of the forces outside of marketing that directly or indirectly influence an organisation's ability to undertake its business. A business has no direct control over the trading forces operating in a marketplace. It is therefore through marketing management that companies build and maintain successful relationships with target customers. The company needs to look at two key perspectives:

- the micro-environment
- the macro-environment.

The micro-environment influences the organisation directly. It includes suppliers that deal directly or indirectly, consumers and customers, and other local stakeholders. 'Micro' relates to a local relationship and it is at this level that the organisation can exercise a degree of influence. Micro areas also include organisation departments, such as management, finance, research and development, and purchasing and operations. Each of these departments has an impact on marketing decisions. For example, research and development have input into a product's performance and accounting approves the financial side of marketing plans and budgets.

The macro-environment includes all factors that can influence an organisation but which are out of their direct control. Influences are continually changing in culture, politics, economics and technology. There may be aggressive competition and rivalry and globalisation means that there is always the threat of substitute products and new competition in the marketplace. For an organisation to survive it needs to be flexible enough to adapt and to remain in touch with its consumers and customers. To achieve this, demography is used to provide information on market segments and target markets.

Demography refers to studying human populations in terms of size, density, location, age, gender, race and occupation. This is very important for marketing managers as it helps divide the population into classifications. This can provide marketers with information about how their product could benefit from a targeted marketing plan. Demography covers other marketing aspects such as family dynamics, geographic shifts, work force changes, and levels of diversity in any given area.

The marketing mix (the four P's of marketing)

The major marketing management decisions can be classified in one of the following four categories:

- product
- price
- place (distribution)
- promotion.

These variables are known as the marketing mix or the four P's of marketing. They are the variables that marketing managers can control in order to best satisfy customers in the target market.

Product—the product is the physical item or service offered to the consumer. In the case of physical products, it also refers to any services or conveniences that are part of the offer. Product decisions include aspects such as function, appearance, packaging, service, warranty, and so on. It is often stated that if a product is good enough it will sell itself by the way designers have tuned into design styles and consumer needs.

Price—pricing decisions should take into account profit margins and the probable pricing response of competitors. Pricing includes not only the list price, but also discounts, financing and other options such as leasing. Multinational hamburger companies never price their products on profit percentages; rather they look at the costs of competing fast food companies and position themselves among them.

Place—place (or placement) decisions are those associated with channels of distribution that serve as the means for getting the product to the target customers. Distribution decisions include market coverage, selecting supply chains and the logistics of delivery, in-store placement and levels of service. Products placed crucially in the public's eye will attract the greatest attention. For example, racks of lollies and lifestyle magazines are always positioned at supermarket checkout lanes to lure the waiting shopper.

Promotion—promotional decisions are those related to communicating and selling to potential consumers. Since these costs can be large in proportion to the product price, a break-even analysis should be performed when making promotion decisions. It is useful to know the value of a customer and the methods of attracting them. Promotion decisions involve advertising, loyalty rewards, public relations and lifestyle creation and are very attractive to consumers who develop an 'emotional loyalty' to a product. Multinational soft drink companies spend millions of dollars to 'sell' a lifestyle as well as a product and at times it can be hard to tell the difference between the two.

Case study: LEGO, designed for business

Figure 7.1 Lego: the building blocks of consumer loyalty

The Danish company LEGO, the world's sixth largest toy maker, realised during a review of market forces that it had to change to meet the demands of trading forces in the marketplace. The changes that the company made have streamlined product development, shaped its business functions and satisfied its customers. This continued improvement in its micro-environment is now being used to improve innovation across the entire business. LEGO, an abbreviation of the Danish words 'leg godt' meaning to 'play well', has remained true to its original mission of producing toys that encourage children to create and use their imagination. LEGO continues to establish itself as an iconic brand among a strong following of users, many of whom have had brand loyalty with the product since childhood.

LEGO has continually evolved its system of bricks and applications. This continual improvement has allowed LEGO to keep up with trends, technology, and ultimately its market share. The extension from LEGO's original system to include moving vehicles and working train sets also set the standard by introducing an advanced kit that included sensors, actuators and programmable logic controllers.

Intense competition has been faced by the company with the explosive growth in computer-based children's toys and the rise of low-cost production in Asia. Further competition between toy retailers is continually pushing margins and prices down. To respond to these market forces LEGO made the decision to reduce the complexity and refocus on the 'classic' product lines. This return to basics received a warm reception from toy retailers and consumers. LEGO also applied a new approach by extensively reorganising and outsourcing many processes and this has transformed the company's financial performance and market share. The idea of LEGO returning to the basics was seen as a decisive move and was used as a competitive marketing weapon.

LEGO chose a variety of data-collection methods such as interviews, surveys and observations in developing the decision-making information. It refocused by using design resources more efficiently and allowed design teams greater participation and responsibility in the collaborative evaluation process. The ideas from the exploration phase were presented formally to the entire project team, and then to key stakeholders, including potential users, their parents, retailers and sector experts. These were then assessed against the objectives set. Feedback in a focus group refined the design recommendations and generated new insights, resulting in a final deliverable set of recommendations on how the products would be taken forward.

7.3 Research methods

The main aim of research is discovering and interpreting as well as developing methods and systems for the advancement of human knowledge. Research is based on the empirical testing of ideas. There are different types and forms of research as outlined below.

7.3.1 Qualitative and quantitative research

Qualitative research features include the following.

- The researcher may know roughly what they are looking for.
- It is recommended during earlier phases of research projects.
- The design emerges as the study unfolds.
- The researcher is the data-gathering instrument.
- Data is in the form of words, pictures or objects.
- Data is subjective: the individual's interpretation of events is important, for example, uses participant observation, in-depth interviews, and so on.
- Qualitative data is 'richer', more time consuming to obtain, and less able to be generalised.

Quantitative research features include the following.

- The aim is to classify features, count them, and construct statistical models in an attempt to explain what is observed.
- The researcher knows in advance what he or she is looking for.
- It is recommended during the latter phases of research projects.
- All aspects of the study are carefully designed before the data is collected.
- The researcher uses tools, such as questionnaires, surveys or interviews to collect numerical data.
- Data is in the form of numbers and statistics.
- Quantitative data is more efficient and able to test hypotheses but may miss contextual detail.
- The researcher tends to remain objectively separated from the subject matter.

7.3.2 Questionnaires

Questionnaires and surveys are an inexpensive way to gather data from a large number of respondents. Every step involved with questionnaires needs to be designed carefully. The steps required to design and administer a questionnaire include:

1. defining the objectives of the questionnaire
2. determining the sampling group
3. writing the questionnaire
4. administering the questionnaire
5. interpreting the results
6. drawing conclusions.

A questionnaire should be considered in the following circumstances:

- when resources and money are limited
- when it is necessary to protect the privacy of the participants
- when corroborating other findings.

Writing a survey/questionnaire

Step 1: Decide what information you want to gather from the survey.

Step 2: Keep the survey as short as possible, asking only those questions that will provide the information you need.

Step 3: Use a casual, conversational style, making the questions easy for almost anyone to understand.

Step 4: Structure the survey so that the questions follow a logical order and evolve from general to specific.

Step 5: Use multiple-choice questions whenever possible. This helps the respondent to better understand the purpose of your question and will reduce the time it takes to complete the questionnaire.

Step 6: Avoid leading questions that might generate false positive responses. For example, the question 'How great was the service provided by our excellent waiters?' should be 'How was the service provided by our waiters?'.

Step 7: Use the same rating scale throughout your survey for questions requiring the respondent to rate items. For example, if the scale is from one to five, with five being the most positive, keep that same scale for all of the questions requiring a rating.

Step 8: Test the survey on ten to fifteen people before you produce it for mass distribution. Conduct an interview with each of those respondents after he or she completes the survey to determine if your questions were easily understood and easy to answer.

7.3.3 Surveys

See 7.3.2 above for information on the use of surveys.

7.3.4 Interviews

The interview method of research typically involves a face-to-face meeting in which a researcher (interviewer) asks an individual a series of questions. When using the interview method you should do the following.

- Prepare interview questions in advance, and share them with the participant so they can give a considered response.
- Tape or video the interview.
- Ask questions if they arise during the interview, even if you did not have them listed.
- After the interview, transcribe exactly what was said: this can be a very time-consuming process.
- Share the written copy of the interview with the participant to make sure that they agree with the content.

Concerns include the following.

- Completeness: was the interview transcribed exactly as recorded?
- Accuracy: did you miss anything?
- Bias: did you 'add' to what you observed by presuming or assuming something that was not stated directly by the participant?
- Clarity: would someone who had not interviewed the participant be able to get a clear, correct picture of what was discussed by reading your notes?
- Confidentiality: did you ask permission for the interview, and is the participant aware of the purpose and intended audience of the interview?

7.3.5 Observation

In observational research the observer does not intervene. The researcher is 'invisible' and does not interrupt the natural dynamics of the situation being investigated. Considerations include the following.

- Use all of your senses, not just your vision. Record the sounds, smells and tastes if applicable.
- Record your impressions and feelings.
- Record the context of the situation: place, time, numbers and choice of participants, and so on.
- Record all of your information in a journal.

Concerns include the following.

- Completeness of information recorded is critical.
- Accuracy of the information recorded is crucial.
- Avoid bias.
- Would someone who had not observed the same thing be able to get a clear, correct picture of what you observed by reading your notes?
- Respect confidentiality.
- Video and audio taping or taking photographs of the situation is infringing on the participant's rights to privacy: use only your written notes.

7.3.6 Tests and experiments

Experimental researchers manipulate variables and randomly assign participants to various conditions. Considerations include the following.

- Experimental research involves defining a research problem, describing a hypothesis, describing the process to be followed, gathering and analysing data, reporting the findings, and stating conclusions in relation to the hypothesis.
- Prepare your experiment in advance: practise your procedure and be sure that you have all of the materials necessary to conduct the experiment.
- Discuss the experiment with your teacher, who will guide and assist you to ensure that the procedure is correct, complete and valid.
- Videotape the experiment: this will help in data analysis, as well as providing some additional information that may be of value when interpreting the data.

Concerns include the following.

- Ethics: is the experiment appropriate, safe, and so on?
- Bias: was anything 'added' to the test results by presuming something?

7.3.7 Statistical analysis

Statistics is the formal science of making effective use of numerical data relating to groups of individuals or experiments. It deals with all aspects of this, including not only the collection, analysis and interpretation of such data, but also the planning of the collection of data in terms of the design of surveys and experiments.

7.3.8 Information research including print and electronic sources

A topical research project involves the acquisition, synthesis, organisation and presentation of information. Considerations include the following.

- Prepare your research questions in advance.
- Consider many different forms of information sources: online websites, paper-based sources such as encyclopaedias, journals, magazines, newspapers, and so on.
- Topical research studies may also include observational research, experiments and tests but other appropriate types of research should also be considered.

Concerns include the following.

- The completeness of information recorded is critical to gain a complete understanding of the topic.
- Because of the variety of information sources be sure that you have reviewed all of the issues for each of the research types.
- Avoid bias.
- Would someone who had not researched the topic be able to get a clear picture of what the topic was all about by reading your report?

7.4 Interpreting and presenting data

7.4.1 Interpreting data

The data or information you initially collect is often in a bulky format (numerical data, transcripts of interviews, descriptions) which needs to be summarised, interpreted and analysed before you can draw conclusions. Summarising information helps to identify patterns. Summarisation enables you to compare information in a standardised format.

After the data is summarised, data analysis proceeds. The data analysis would first focus on the specific research hypotheses, to see if the hypotheses or predictions are supported by the data. An analyst will also explore the data to see if anything unexpected or unanticipated emerges. After the data analysis is complete the results are written up in report form.

7.4.2 Presenting data

You need to present your information clearly. The use of figures (diagrams, photographs, maps, graphs, and so on) or tables (lists of written or numerical information) will save you from writing lengthy descriptions and enable you to demonstrate the crux of your arguments. Remember that all figures and tables must:

- be numbered consecutively
- be correctly referred to (by number) and relevant to the text
- be presented in a consistent style
- have a descriptive caption so that they can be understood without text if necessary.

When planning your data presentation:

- focus on your message
- decide on the target audience
- decide how the message will be delivered
- pick your time and place
- inform, educate and persuade!

7.5 Ethics in research

It is important to adhere to ethical norms in research because these promote the aims of knowledge, truth and avoidance of error. Research often involves a great deal of cooperation and coordination among many different people in different disciplines and institutions. This requires ethical standards which promote the values of trust, accountability, mutual respect and fairness. Most researchers want to receive credit for their contributions and do not want to have their ideas stolen or disclosed prematurely. Many ethical norms also help to ensure that researchers can be held accountable to the public.

Ethical norms also help to build public support for research. People are more likely to fund a research project if they can trust in its quality and integrity. Many of the norms of research promote a variety of other important moral and social values such as social responsibility, human rights, animal welfare, compliance with the law, and health and safety. Ethical lapses in research can significantly harm humans and animals.

The principle of voluntary participation requires that people not be coerced into participating in research and that there is informed consent. This means that prospective research participants must be fully informed about the procedures and risks involved and must give their consent to participate. Ethical standards also require that researchers not put participants in a situation where they might be at risk of harm as a result of their participation. 'Harm' can be defined as both physical and psychological.

Almost all research guarantees the participants' confidentiality in that identifying information will not be made available to anyone who is not directly involved in the study. The stricter standard is the principle of anonymity, which essentially means that the participant will remain anonymous throughout the study, even to the researchers themselves.

The main issues to consider in relation to ethics in research are listed below.

- Honesty: strive for honesty in all scientific communications including data, results, methods, procedures and in publications.
- Objectivity: strive to avoid bias in experimental design, data analysis and interpretation, peer review, personnel decisions, grant writing, expert testimony, and other aspects of research where objectivity is expected or required.
- Integrity: keep your promises and agreements; act with sincerity and strive for consistency of thought and action.
- Carefulness: avoid careless errors and negligence: carefully and critically examine your own work and the work of your peers and keep good records of research activities.
- Openness: share data, results, ideas, tools, resources, and so on and be open to criticism and new ideas.
- Respect for intellectual property: honour patents, copyrights, and other forms of intellectual property.
- Confidentiality: protect confidential communications such as papers or grants

submitted for publication, personnel records, trade or military secrets and patient records.

- Respect your colleagues and treat them fairly.
- Social responsibility: strive to promote social good and prevent or mitigate social harms through research, public education and advocacy.
- Non-discrimination: avoid discrimination against colleagues or students on the basis of sex, race, ethnicity or other factors that are not related to their competence and integrity.
- Competence: maintain and improve your own professional competence and expertise through lifelong education and learning.
- Legality: know and obey relevant laws and institutional and governmental policies.

Key concepts and definitions

Aesthetics—when designers build in features that are visually important or attractive.

Consideration involves design aspects that require further thought or research.

Constraint involves design aspects that are fixed and will not be changed.

Design brief—intended to clarify the need by developing a common understanding of what is to be accomplished. It will continually provide direction by identifying a course of action while clearly communicating intentions to others involved.

Design process—the planning and executing of an idea into a solution. It generally involves a sequence of pre-planned activities that develop problem-solving skills and creativity.

Function—the project's ability to serve a purpose.

Life cycle analysis—undertaken by designers to assess all the steps and implications of a product's existence. The sum of all those steps or phases is the life cycle of the product.

Marketing mix or the four P's of marketing—the variables that marketing managers can control in order to best satisfy customers in the target market.

Obsolescence—when particular items become obsolete as they reduce their function due to natural wear, or when a new product or technology becomes preferred.

Primary research—the use of immediate or up-to-date data.

Prototype—an original, full-scale and usually working model of a new product, or new version of an existing product.

Qualitative research is largely determined by the respondents' own thoughts and feelings.

Quantitative research deals with numbers and often involves statistical analysis.

Secondary research—a means of reprocessing and reusing collected information to improve a service or product.

Useful websites

- freelanceswitch.com/clients/the-ultimate-design-brief/
- www.lego.com/en-US/default.aspx
- www.marketingteacher.com/Lessons/lesson_marketing_research.htm
- www.netmba.com/marketing/mix/
- www.p3.unsw.edu.au/Forms/The_Design_Brief.pdf
- www.rebelsport.com.au/ecom/rebel/default.aspx

Classroom activities

1. You have been requested by a corporate client to design a new pair of cross-training shoes. In the T-chart below identify all the design parameters listed as constraints and considerations for the product to meet the need.

Constraint	Consideration
Fixed and will not be changed	Requires further thought or research

2. Using an A3 page, develop a series of concept sketches for the new cross-training shoes. Identify each new design element employed by incorporating ancillary sketches around the page. Experiment with colour combinations.
3. Identify the types of research techniques that would be incorporated when conducting market research for a new line of cross-trainers.

4. Survey a variety of student groups to determine the branding, style and purpose of various shoes used in sport.
5. Design a package for the cross-trainers. Include corporate colours, logo, an image of a sportsperson promoting or using the cross-trainers, and so on. Use text appropriately to identify the product and provide an explanation of its style.

Key concept questions A pp. 202–203

1. Explain the importance of the design brief when working towards a successful MDP.
2. Identify which forms of communication would be most suitable in the early stages of the design process in order to demonstrate to a client how a new design idea would work.
3. Describe four factors of design and identify their appropriateness when designing a new pair of cross-trainers.
4. Explain the importance of developing criteria to establish success.
5. Companies such as Rebel Sports use sophisticated marketing management to build and maintain successful customer relationships. List the factors that would influence both the micro- and macro-environments of this sporting goods outlet.
6. Outline how the marketing mix could be used by Rebel Sports to best satisfy its target customers.
7. Explain why packaging, branding, warranty and service are included as part of the product in the marketing mix.
8. How does a company the size of Rebel Sports responds to the market forces that drive competition?

Sample Preliminary questions

Objective-response question A pp. 203–204

Circle A, B, C or D. (1 mark)

1. In any design process, successful designing uses research to

A provide a market for innovations.

B help keep costs down.

C maintain long-term profitability.

D maintain positive decisions throughout the process.

Short-answer question A p. 204

2. Merrell, a leading footwear company, is planning a new range of hiking boots. Describe what market research methods would be used to collect, analyse and evaluate data about this particular target market.

(5 marks)

8 – Effective use of resources

PRELIMINARY OUTCOMES

A student:	You learn about:	You learn to:
P4.2 uses resources effectively and safely in the development and production of design solutions.	– using materials, tools, techniques and other resources – characteristics and properties – functions and uses – experimentation – criteria for selection – consequences of use ■ the realisation of ideas through the manipulation of materials, tools and techniques and other resources. ■ safety – safety in the use of materials, tools and techniques – legislative requirements including Occupational Health and Safety.	■ select appropriate materials, tools, techniques and other resources. ■ justify and explain the selection and use of resources in design projects. ■ develop and demonstrate proficiency in using an appropriate range of materials, tools, techniques and other resources. ■ implement safe work practices when designing and producing.

8.1 Uses resources effectively and safely in the development and production of design solutions

8.1.1 Using materials, tools, techniques and other resources

Products are becoming more complex as machines substitute labour and computing accelerates the interaction of each process. Technology has the ability to rapidly magnify human capabilities and the potential for creating many types of improvements. As manufactured products become more complex in their requirements, the need to design and produce them in time to a high standard has also increased. The response as a designer is to maintain control over the complex tasks by deliberate choices in materials selection, strong control over the tools and processes used, maintenance of quality and control over safety throughout. As a designer you need to determine your goals and build your capabilities as well as recognise and include the factors that will aid you in selecting and using resources effectively and safely.

Developing a design brief will help you to collect your ideas and identify ways that the design may develop. From here you will begin to use your knowledge, skills and resources to make or change things. This is an important step in designing and producing and as the design process proceeds the brief will help you find solutions in a step-by-step manner. A great deal of a designer's time is spent on research of the properties of materials and tools in relation to their practical use. By investigating and comparing the properties of different materials, for example, paper, card, wood, textiles, metals and polymers, a designer can select the best for the particular design.

When choosing resources appropriately to make a sports bag or spray jacket, a designer would decide on a material that was waterproof. This would be the result of a testing regime that identifies the design's ability to resist water. Once the material is selected it would then be a process of designing appropriate water-resistant seams and fasteners.

8.1.2 Characteristics and properties

Designers employ a series of processes to determine what material will best suit the design application. This information can be found by material testing, comparing and experimenting. Using published technical data can also be of some use. It may also involve a collection of data mining tools including primary and secondary research techniques. Many designers also engage the advice of experts while continuing to explore redesign options by studying existing ideas. Understanding the purpose of the materials to be used and their related properties is important in developing strong features in the design.

8.1.3 Functions and uses

The functionality of design is highly regarded by the designer. Based directly on the design specifications, function will determine the design's performance. Specific design elements that impact on the materials, tools and manufacturing need to complement each other if the design is to be successful. There are times when design becomes the fashion and a designer tries too hard to leave a 'design' imprint. This generally creates what is termed as 'design excess'. The resulting design becomes very clever but generally loses out on functionality. It almost seems a contradiction since design is meant to aid function.

Designers need to be aware of these built-in design complexities and avoid them at all costs. A designer may design a truly beautiful drinking vessel but it may be almost impossible to wash or even to hold. A sleek and elegant car that looks great in photographs may be very awkward to drive. An object will be best remembered by its functional design maximising its successful use.

8.1.4 Experimentation

To discover what works there are times in a designer's research where they will resort to experimentation. This is generally when there are no secondary research results available and/or when a designer is trying something new or unproven. Essential to this is the reliability of the results, which necessitates a tightly-controlled environment. In every experiment consistency, predictability and reliability must be assured.

Experimentation identifies variations and provides a platform to make appropriate suggestions for changes that would lead to an improved outcome and the use of innovative alternatives. Designers relate their findings to the purpose for which the product and/or system was designed and the appropriate and ethical use of the resources. Most experimentation is carried out to meet particular requirements in the design brief.

8.1.5 Criteria for selection

In developing suitable criteria designers select from the vast range of materials, tools and techniques available. They take account of function and performance, energy requirements, aesthetics, finances and ethical and legal considerations. Designers also base their decisions on the understanding of the properties of materials or ingredients. By developing a range of criteria designers are able to make decisions that will address the requirements of the design brief.

8.1.6 Consequences of use

While design elements can be isolated in any form, designers prefer to apply these together to form a successful balance. These elements can carry a wide variety of meanings to complement a strong, well functioning design. If this was to alter due to changing circumstances the designer would adapt their design processes to determine what is best suited for the new application. The designer needs to be flexible in responding to changes in consumer needs and variations in the market. When materials selection, tools and manufacturing do not complement the design to its end use it can have catastrophic consequences. To suppress the impact of public perceptions of design failure many products are recalled. Distributors and retailers remove them from sale and customers are advised of the situation through advertising in a wide range of media.

A product recall was initiated on a drink bottle. Even though the materials were well selected the design failed its function as the pop-top device could be removed, thus presenting a potential choking hazard for children.

A chemical used in hard plastics and in the linings of food and beverage cans, bisphenol A (BPA), is linked to health problems including diabetes and heart disease. When people drank from bottles containing BPA the chemical leached out into their bodies. Health authorities in the USA urged American consumers to lower their exposure levels by buying products in glass, consume frozen or fresh foods, and use powder-formed foods rather than packaged liquid foods. Another problem was the safe disposal of the product. The material selection did not meet expectations, causing the failure of the product. Designers must understand the relationships between materials/ingredients, systems components, techniques and sustainable design.

8.2 The realisation of ideas through the manipulation of materials, tools and techniques and other resources

The work of a designer is planned and organised in such a way that they can carry out processes accurately, consistently and with precision. It generally starts with design sketching where all ideas are brought out quickly on paper. Further sketching will produce better detailed ideas that may be applied to collaboratively communicate ideas with others.

The designer may then use a variety of 3-D media such as cardboard, clay or foam to make models. Information gained here can be used in other areas of the design process. This evaluation helps towards achieving working drawings, and the eventual production of a prototype.

8.3 Safety

8.3.1 Safety in the use of materials, tools and techniques

Designers continuously push the boundaries of design manufacturing. They increasingly choose complex production techniques and equipment in an effort to produce 'cutting edge' results. They acknowledge the importance of the safe use of tools and equipment and organise their working environment so that the materials, equipment and space safely support the particular activity. This safe and confident use of materials and tools is reflected in their training and supervision and their understanding of the risk assessment process. Design and Technology students also need to approach each task in a competent and safe way. Having knowledge of the properties of different materials and the processes that can be safely applied to them is very important.

To be at a level where production techniques and equipment can be used in a safe and confident manner, students must undergo training and supervision. This will help them develop an understanding of the properties of materials and tools in relation to their practical use. Applying a risk assessment to each process will provide information on the safe handling of materials and equipment, the correct protective clothing, and how and when to ask for help or report faults.

While it is important to develop a strong safety ethos in all your work activities it is also important for companies and institutions to provide you with a safe working environment. Companies are required by law to ensure safe working environments. The law, as outlined in legislation, is administered in Australia by WorkCover. Once a regulation is law, all employers and employees must meet their safety obligations under the legislation. If anyone fails to meet these regulations they are liable and this can lead to a company being fined, shut down or, in extreme cases, individuals jailed.

The *Occupational Health & Safety Act 2000 OH&S Act 2000* is relevant to all NSW workplaces. The Act supports requirements necessary to ensure a safe and healthy work environment and is designed to reduce the number of workplace injuries by imposing responsibilities on individuals and corporations. The *OH&S Act 2000* requires employers to consult with workers on matters affecting their health, safety and welfare. The Act ensures that risks are identified and assessed and then eliminated or controlled. The Act takes in progressively higher standards of health and safety to take account of changes in technology and resulting work practices. The Act also involves public safety and it promotes community awareness of occupational health and safety issues.

The expected and/or required standards are contained in an industry code of practice. Codes of practice provide practical guidance and advice on how to achieve the standards required by the Act. Australian standards set out the safety requirements and provide guidance for persons working in specific areas or who deal with particular equipment. These standards only become legally binding when they are incorporated into legislation. Codes of practice are developed through consultation with representatives from industry, workers and employers, special interest groups and government agencies.

Managers and supervisors are directly responsible for OH&S within areas under their control. To ensure they meet their responsibilities, the code of practice would ensure that they provide:

- safe premises
- safe machinery and substances
- safe systems of work
- provision of health and safety information, instruction, training and supervision
- suitable working environment and facilities
- for the safety of people other than employees who may be present at the workplace.

A section of the Act outlines some key responsibilities for workers. A worker must take reasonable care for the health and safety of any co-workers who may be affected by their actions, and each worker must cooperate with the employer in ensuring a safe workplace. For workers to meet their responsibilities the code of practice would ensure that:

- their actions do not put others at risk
- they are carrying out work in a safe manner
- they use and maintain machinery and equipment properly
- their work area is free of hazards
- they notify their supervisor of actual and potential hazards
- they wear or use prescribed safety equipment
- they follow health and safety instructions
- they take notice of signs
- they participate in safety training.

Workers are independently responsible for health and safety in their area, including the identification and control of hazards. They are required to respond to:

- identifying hazards
- assessing potential risk posed by the hazard
- controlling the risk
- reviewing the risk assessment.

A risk management process involves each party obtaining and providing relevant and adequate information when required. The *Occupational Health & Safety Regulation 2001* requires manufacturers to prepare material safety data sheets (MSDS) before a hazardous substance is supplied to another person for use at work. The MSDS must include the following information:

- recommended uses.
- chemical and physical properties.
- description of each ingredient used in preparation.
- relevant health hazard information.
- precautions to be followed for safe use.
- emergency contacts and general contact details of the manufacturer.

There are many stories of courage and determination following on from a devastating accident at work. WorkCover works towards rehabilitation and spreads a strong message about workplace safety, injury prevention and the personal journey required for recovery and an eventual return to work.

Key concepts and definitions

Design excess—when a designer seeks too hard to leave a design imprint and therefore 'over-designs' the product, usually compromising function in the process.

Industry codes of practice provide practical guidance and advice on how to achieve the standard required by the OH&S Act and Regulation.

Legislation—the regulations detailing the requirements that companies must observe. If anyone fails to meet these they are liable.

The ***Occupational Health & Safety Act 2000*** is relevant to all NSW workplaces. It describes the general requirements needed to ensure a safe and healthy workplace.

Risk assessment—an estimate of the likelihood of adverse effects that may result from exposure to certain health hazards, especially pollutants in the environment.

Useful websites

- whatthehellcanieat.wordpress.com/
- www.dpc.nsw.gov.au/__data/assets/pdf_file/0006/10896/Dignity_Respect_Charter.pdf
- www.nswbusinesschamber.com.au/?content=/channels/Occupational_Health_and_Safety/OHS_ Law/_Overview/how_to_comply_with_ohs_law.xml
- www.ohs.com.au/Sites/
- www.workcover.nsw.gov.au/Pages/default.aspx
- www.workcover.nsw.gov.au/Publications/OHS/OHSResponsibilities/Pages/workcover_watching_out_for_you_poster.aspx

Classroom activities

1. Choosing characteristics and properties appropriately is important to the designer. List the characteristics and properties required for the identified designs below. (See table below.)
2. 'Google' 'product recall' and look at the latest recalls and withdrawals. Identify how many of the products may have endangered our health, and which products failed with the product disclosures on their packaging.
3. Go to the WorkCover website and list the roles of WorkCover in enforcing OH&S standards.
4. Design a poster that reinforces safe activity. To gain ideas of appropriate activity 'Google':
   - Don't turn a night out into a nightmare.
   - 'WorkCover watching out for you' poster.
   - Stop, Revive, Survive.
5. Identify the members on the OH&S committee at your school.

Design	Characteristics and properties
Wetsuit	Thermal, seam fixture, fasteners, fitting and colour
Umbrella	
Racing-bike helmet	
Tent	
Bookshelf	
Metal jewellery	
Snowboard	
Jeans	

Key concept questions A p. 204

1. Identify the series of research processes that designers employ to determine what material will best suit the design application.
2. Explain what is meant by the term 'design excess'.
3. When experimenting, describe why a designer needs to provide a very tightly-controlled environment.
4. List some of the considerations used by designers and explain how they are used to develop suitable criteria for designing.
5. List a range of strategies designers would use when they are developing appropriate ideas for their design decisions.
6. State some of the methods designers employ to ensure safety when they are working with materials, tools and processes.

Sample Preliminary questions

Objective-response question A pp. 204–205

Circle A, B, C or D. (1 mark)

1. When selecting resources appropriately for a new project, a designer needs to consider

A the results obtained from analysis of research.

B the location of the resources required.

C the level of the designer's skill and ability.

D the suitability and availability of tools to the preferred material.

Short-answer question A p. 205

2. The stimulus material below highlights the fact that designers are able to use a wide range of varied sources of materials to develop their ideas into fully realistic designs. Explain how varied materials are applied by designers as part of planning their design decisions. (5 marks)

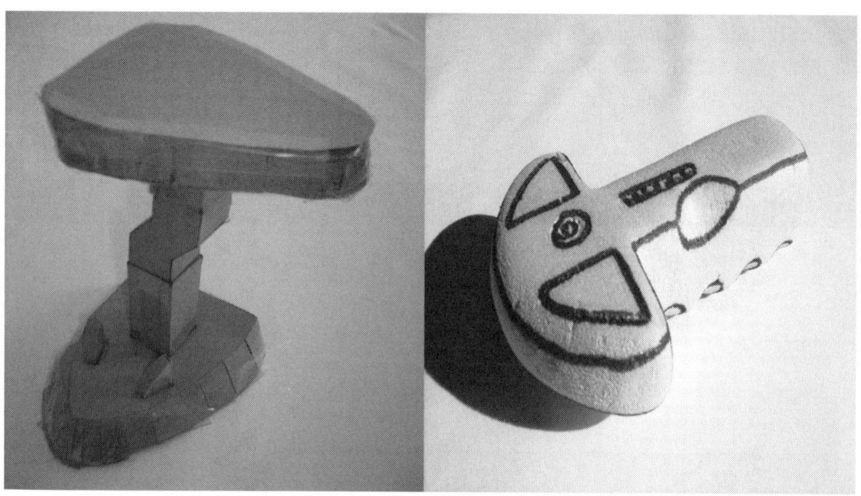

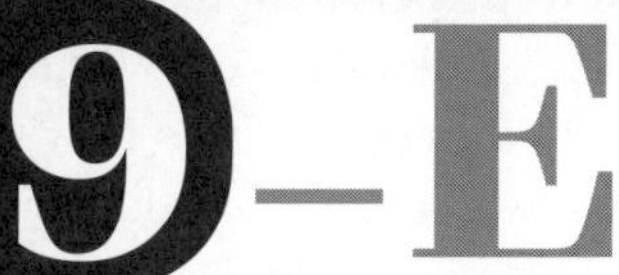

valuate designing and producing

PRELIMINARY OUTCOMES

A student:	You learn about:	You learn to:
P4.3 evaluates the processes and outcomes of designing and producing.	■ evaluation – developing and refining ideas – criteria for evaluation – methods of evaluation.	■ establish the parameters for design and identify criteria for the evaluation of design projects. ■ examine processes undertaken in design projects. ■ conduct continual evaluation throughout design and production. ■ assess the impact of designing and design projects on society and the environment. ■ test and evaluate the appropriateness of design solutions.

9.1 Evaluation

Some designers are so convinced that their design solution is the best possible that they move through the design process too quickly and risk failure. They may fail to recognise that their design was not finally evaluated and therefore might never correctly meet its need. If this occurs, the designer must try to find an opportunity to correct it. This usually comes at a great cost due to unnecessary spending and/or lost time. For a product that fails, it may mean a product recall or withdrawal resulting in lost sales and market share. The fastest way designs fail is when the introduced idea never completely delivers the promised features or benefits.

People who buy, then reject a product are unlikely to purchase again, despite the improved design. This can have a devastating effect on the company's reputation. Designers and Design and Technology students may never realise the success of their design without a conscious plan of evaluation. According to Philip Kotler in *Marketing Management*, it is estimated that over forty per cent of newly packaged consumer products fail, some twenty per cent of new industrial products fail and eighteen per cent of new service products fail. Whether it is big business or small design, a conscious plan of evaluation is always necessary.Evaluation is ongoing and continuous!

9.1.1 Developing and refining ideas

Through a series of decision-making processes when developing and refining ideas, a designer recognises the possibility of the design as an innovation by developing ideas for something that either did not exist or by improving a current design. Throughout the course of the design process the designer is continually checking their developing ideas against criteria of success. This is a way to keep the design flowing on track and helps in refining the idea. It is this continual process of refinement that produces the 'right' design.

9.1.2 Criteria for evaluation

The most successful evaluations are those that are measurable, definable and actionable. The criteria that will be used needs to be carefully selected as it will be what is used to judge the success of the final design. Many of the design decisions, strategies and specific requirements will be based on the criteria of evaluation, in a similar way to how major companies define a mission statement.

Consider a famous refrigerator manufacturer who had sales growth only at the rate of new home-building so who undertook a process to define their purpose: were they a company that built refrigerators or was their greater purpose to preserve food? They decided they were in the business of food preservation and consequently moved towards new product ventures. For them it was a call to action.

As a Design and Technology student it is important that you define the purpose with specific goals and put your ideas into action. This can be accomplished via the selected criteria. To define suitable criteria you need to recognise how the successful design will look at the end. You then draw in the elements that link between present and future to come up with a clear vision of the successful design. Well-detailed criteria will provide strong vision and direction but there may also be times when the criteria must be revised to ensure it is still appropriate, significant and relevant. It always needs to be matched against your design brief.

In industry, many companies use a Functional Design Specification (FDS). It is a document produced in the pre-development phase to identify all the concepts within the scope of the design. The document can include flowcharts, screenshots and decision-making charts. At a minimum, an FDS will contain an organised list of criteria that can be used for development, testing and client sign-off. The benefits of an FDS can include the saving of time, resources and money at the designing, developing, testing, and approving stages. As one designer pointed out, an FDS prevents the headache of trying to translate sticky notes, ideas written on cocktail napkins and countless emails.

In writing criteria to establish success you need to deal with specific issues relating to the materials, tools, level of technology and how the final design fits in the broad scheme of things. You need to be aware of how your design will satisfy a need, solve a problem or provide support. There may be special challenges such as positioning strategies in the marketplace and using measurable tests to benchmark your design against competing ones.

Case study 1: mobile phone dual-SIM touchscreen

Most mobile phones now contain a range of media functions. These can include MP3 and MP4 players, digital cameras and image viewers, and FM radio. Touchscreen media phones also include alarm clocks and task management software with well-designed extras such as Bluetooth and USB connectivity. Many mobile phones now have two SIM slots with both call waiting and call divert options for each SIM card. It provides the flexibility of having a work and personal number on the same handset. While dual SIM phones are nothing new in Asia they have taken longer to appear in Australia due to the protocol of sharing one device between two mobile phone companies.

Function as criteria

The criteria used to evaluate the functional success of a mobile phone could be based on the following features:

- caters for a variety of user groups.
- two SIM slot capability.
- provides a variety of features.
- data volume.
- has clear onscreen instructions.
- simple pathways to features.
- versatile language base.
- user-friendly controller system.
- warranty.
- acceptable size.

Aesthetics as criteria

The criteria used to evaluate the aesthetic success of a mobile phone could be based on the following features:

- variety of colours.
- styles that appeal to the market.
- good lines and efficient shape.
- a catalogue of stored sounds.
- textural feeling in the grip and ergonomic to hold.
- consistent page display.
- interesting and appealing text characters.
- strong use of graphic displays.

Case study 2: Strung-out clothes rack by F3 Design

A neat idea from New Zealand is a clothes rack that is ceiling mounted and can be wound so that the hanging washing is above head height. This industrial-styled design is made from galvanised steel and is connected with marine-style non-corrosive fixtures.

For many products designed for their utility it is often true that their functional requirements will far outweigh their aesthetic value. Some would argue that a utility object may not need to be

Strung-out clothes rack

Criteria to evaluate success		Evaluation decisions
Function	■ Ability to be out of the way yet accessible ■ Handles the weight of wet washing ■ Handles the contents of a large-capacity washing machine ■ Will not flex under load ■ No stains on the washing ■ Built from a 'flat pack'	■ To be affixed to ceiling ■ Ease of use, incorporate a mechanical advantage ■ Large line capacity ■ Strong yet lightweight tubular frame ■ A non-corrosive and easily cleaned finished surface ■ Ease of assembly
Aesthetics	■ Utility colour to match outdoors ■ Colour match to frame ■ Neat appearance when clothes drying ■ Sleek appearance ■ Reduces the look of 'clutter' ■ Possibility of coming in a variety of different colours	■ Industrial metal colour ■ Black connectors ■ String lines set evenly apart ■ Smooth texture ■ Level with the line of ceiling ■ Use of a variety of finishing processes

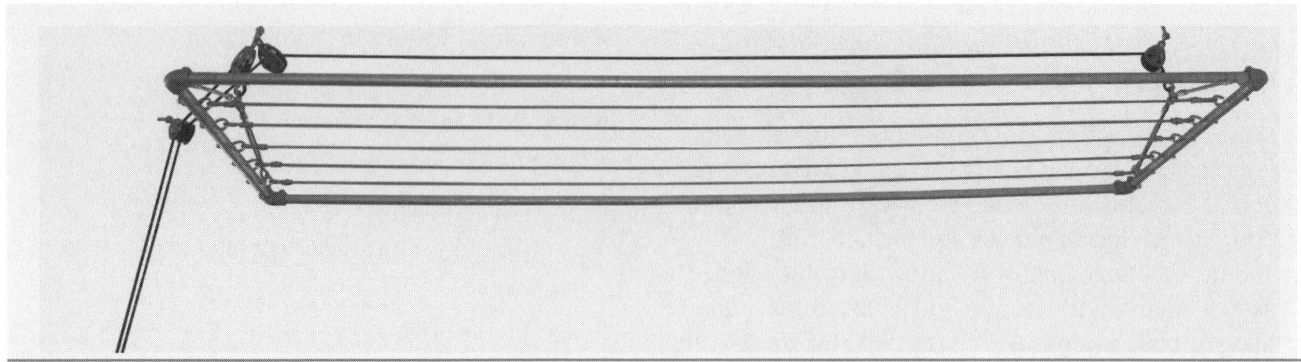

Figure 9.1 There is nothing 'strung out' about this design which encompasses a true balance of aesthetics and function.

Source: © F3 Design 2009, www.f3design.co.nz, Christchurch, New Zealand

aesthetic at all. The design philosophy named 'user-centred design' gives extensive attention to the needs and wants of the end user at each stage of the design process. It recognises that design aesthetics may be utilised to apply a theme or style to the object as long as this does not compromise usability. It is the responsibility of the designer to ensure that it is well designed functionally and that it also has a strong aesthetic value. It is the aesthetics that develops basic human intimacy between the user and the product.

9.1.3 Methods of evaluation

Most people have an intuitive appreciation for what success is but defining and measuring it may not be so easy. To measure the success of the final design it will need to be judged against a set of parameters based around the functional and aesthetic design factors. Whether the design is ultimately successful or not, you will be able to identify its strengths and weaknesses.

Most design projects will have a list of specifications that measure the success based around the satisfaction of the final result. When developing the criteria to establish success you need to ensure that the specifications you plan to measure are important and realistic, that they cover the expected design specification, and are comparable with existing criteria.

A common theme in the construction industry is that the success of a building may be measured in the time it took to complete and its final costing. If these criteria were to be used in the case of the Sydney Opera House then it may be specified as a complete failure. It cost sixteen times the estimate and took four times longer to complete than the original plan. While it is recognised in the construction industry as a project-management disaster, it produced an enduring and inspiring civic symbol.

The varying types of designs will call upon varied evaluation tools. Some of the methods employed by designers would include:

- usage and attitude studies that examine consumer use
- controlled field testing of the physical object
- low-grade qualitative research based around talking to consumers or surveying about product satisfaction.

The selection of the evaluation methods needs to be directly related to the end use of the design. As with any other technique, project success measures can be overdone. To avoid this you should ensure that all criteria are:

- valid—they measure what you intended to measure
- valuable—you understand the value of the information
- relative difference—before and after comparisons to the results are measured
- complete—anything unmeasured is likely to be compromised
- easy to understand
- relevant—any variation may indicate a need for adjustment or redesign.

The only necessary requirement after evaluation is done is to document all the proceedings and the results. Conclusions can then be drawn from your evaluations.

Design project example: a cross-country skiers' backpack

You are planning the design of a backpack to suit your individual needs. While there are commercially produced backpacks, none suit your requirements for cross-country skiing. You will be able to incorporate the ideas already employed, including:

- a padded shoulder area
- a removable padded shoulder strap
- mesh 1000 ml water bottle pocket
- gear attachment sites
- bottom compression straps to secure a sleeping bag or jacket
- an organiser with mobile phone pocket and removable key fob
- a zippered top pocket, with headphone port, holds a CD/MP3 player or sunglasses.

Some of your individual needs to innovate the existing product will require the following:

- a main compartment with room to stow cross-country ski boots
- positioned Velcro ties for ski poles
- a hydration unit with insulated water tube to replace the water bottle
- alternative gear attachment locations to include prominence of a GPS navigational aid
- an identifiable section for a first-aid kit
- the reduced look of 'clutter'
- a sleek appearance in a variety of colours so that the aesthetics and functionality complement each other.

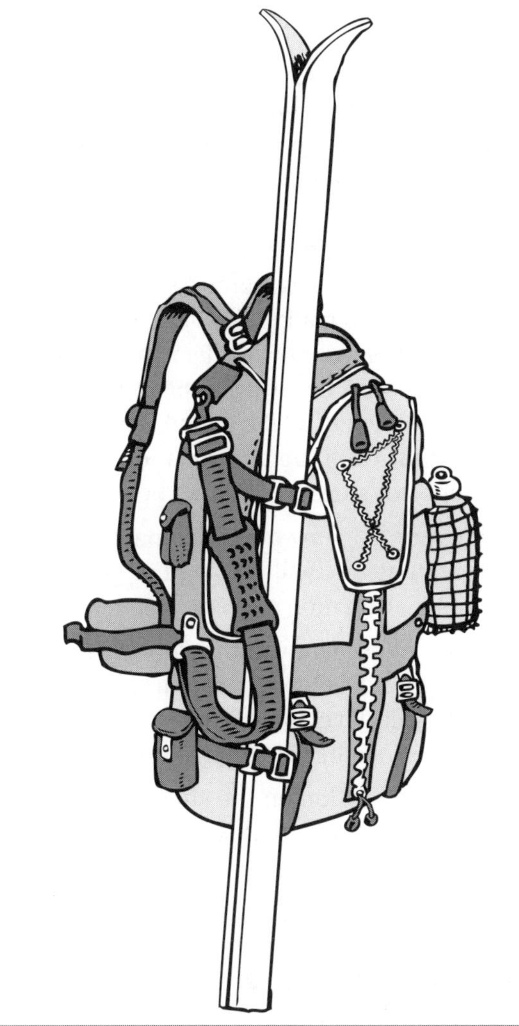

Figure 9.2 A cross-country backpack to suit skiers' needs

You now set out to develop testing methodology to assess the success.

Cross-country skiers' backpack		
Criteria to evaluate success		**Testing methodology**
Function	■ Main compartment with room to stow cross-country ski boots ■ Positioned Velcro ties for ski poles ■ Hydration unit with insulated water tube to replace the water bottle ■ Alternative gear attachment locations to include prominence of a GPS ■ Identifiable section for a first-aid kit	■ Measurement ensuring stowage of different shaped and sized boots ■ Ski poles ergonomically positioned ■ Insulation test of the tube in a freezer ■ Identify suitable location for viewing GPS ■ Identify position of first-aid logo on an easily seen and accessed site
Aesthetics	■ Sleek appearance ■ Reduced the look of 'clutter' ■ Possibility of coming in a variety of different colours for fashion	■ Comparison test on a variety of water-resistant textiles ■ Compare zipper-styled compartments ■ Colour selections for style ■ Use of survey in an outdoor store

Figure 9.3 An innovation to the backpack to suit skiers' needs

Continuous evaluation

While some designers see the end result as the most important factor, the philosophy of 'getting there' also requires some attention. After all, if the design path is not planned with continuous evaluation then the design may never reach its 'destiny'. The advantage of continuous evaluation is that it removes all of the uncertainties of the design process so you are able to determine the right path. It also develops a consistent approach, allowing you to decide on and control the feedback. This will ensure the smooth processing and better execution of all the design stages. Planning and conducting continuous evaluation requires an ongoing series of steps as outlined below.

- Identify your goals by developing a strong design brief as this will help you to guide the continuous evaluation process.
- Recognise where you want to go with the designing and develop a course to get there. As the project develops you will recognise how activities fit together in coherent ways.
- Set a clear indication on the short-term, interim, and long-term results that will measure your achievements.
- Set up a system for managing the information you need to measure. In this way your evaluation will lead you straight to the results.
- Analyse your results and use what you learn from them. Having the right information available at the right time will allow you to make the right decisions.
- Be patient. Although you cannot expect dramatic results immediately, take heart that you are making the right decisions as you progress, and that each stage will build on to the next.

The impact of designing and design projects on society and the environment

Many designers believe in the power of design thinking to contribute to significant social change. They engage in projects that enable communities, individuals and economies to grow through thoughtful design approaches. They recognise the importance of social responsibility in design's human-centred approach. Designers have the responsibility of designing in such a way as to reduce the negative impacts on our society and environment and work towards sustainable futures. Many of the impacts, however, are not fully realised until full testing is done and in some instances the impact may occur in use. Design is not neutral and it is important for designers to weigh up both positive and negative factors when assessing any design for its social and environmental impact.

In assessing the impact of a design on society issues such as safety, ease of use and obsolescence need to be considered. The designer needs to question and understand how the design could improve standards of living, how it would improve equity, and how it might even support a higher moral purpose. Some groups within society such as the elderly and disabled often have needs that require different design approaches and methods.

In assessing the impact of the design on the environment the designer should consider the effects it will have on the sustainability of resources, and all the environmental considerations associated with manufacture and the disposal of the end product.

Keeping society and the environment in balance

Some domestic smoke detectors contain a quantity of Americium-241, a radioactive material that causes concerns about disposal as the detectors take years to break down in landfill sites. Designers must balance this environmental impact with the life-saving benefits of smoke detectors. Smoke detectors used for commercial or industrial purposes which contain excess Americium-241 are required by law to be returned to the seller for disposal. This is one of the ethical responsibilities considered by the designers, and with help from government regulatory bodies they can ensure sales companies accept the unwanted detectors for safe disposal.

Balancing society's needs with responsible design

As Australia has the highest rate of skin cancer in the world it is important that Australians protect themselves with sunscreens containing zinc oxide and titanium. The iconic 'white swipe' across the face is to be replaced with new nanoparticle technology that will decrease the particle size and not be seen on the skin. A theoretical concern has been raised that the absorption of the nanoparticles into skin cells could possibly produce free radicals, damage DNA and be toxic to cells, especially when exposed to UV light. Currently nano-sunscreens are not labeled, making it difficult to make an informed choice. Due to strong public support there are calls for mandatory product labeling, which can also be seen as a step towards developing labeling on cosmetics and other personal care products. Around 1600 Australians die of skin cancer each year and the ethical responsibility for any designer working for a pharmaceutical company is to provide skin protection without the risks.

Key concepts and definitions

Bluetooth—a short-range radio technology for Internet and mobile devices, aimed at simplifying communication among them.

Criteria to evaluate success establishes criteria which clearly state the most important functional and aesthetic features of the intended design.

Evaluation—the process of careful examining and judging. The evaluation process can often lead or guide further actions.

Functional Design Specification (FDS)—
a document produced in the pre-development phase to gauge the success of the design.

Useful websites

- eco-ideas.net
- gsociology.icaap.org/methods/
- managementhelp.org/misc/designing-eval-assess.pdf
- www.hsc.csu.edu.au/design_technology/producing/projadv/2-0/2.0.6.html
- www.ibm.com/developerworks/rational/library/2950.html
- www.japanfs.org/en/mailmagazine/newsletter/pages/027866.html
- www.smartdesignworldwide.com/thinking
- www.upassoc.org/usability_resources/about_usability/what_is_ucd.html

Classroom activities

1. Brainstorm in small groups and collect experiences of 'shoddy' merchandising such as the breakdown of physical components, the running of dyes or the ripping of fabrics. Share the material gathered with other groups.
2. Go to www.viewpoints.com/Fisher-Price-See-n-Say-Elmos-Peek-amp-Seek-review-77703 and read the first-hand account of a parent who experienced her child choking on a toy. Develop a plan to resolve this problem for consumers of the toy.
3. Read the material below and conduct a Functional Design Specification (FDS) to document the redesign of one of the faulty toys for redistribution:

   Toy company Mattel recalled nine million more Chinese-made toys, including Barbie, Polly Pocket and *Cars* movie items, and warned that more could follow because of lead paint and small magnets that could be a choking hazard. Mattel had recently recalled 1.5 million Fisher-Price infant toys, also made in China, because of potential lead paint hazards, and there was a government warning to parents to keep the recalled toys away from children.
4. Use a table to identify the criteria needed to evaluate the success of a set of car seat covers. (See below.)
5. Make a list of products that you feel have been made responsibly. Identify the attributes that make these designs possible. Examine each attribute for similarities and differences.

Key concept questions A pp. 205–206

1. When designing, why is it necessary to develop a conscious plan of evaluation?
2. Why is a continuous plan of evaluation equally important to a final product evaluation?

Design project: a set of car seat covers		
Criteria to evaluate success		**Testing methodology**
Function		
Aesthetics		

3. What are functional and aesthetic criteria?
4. Analyse the design of a digital camera in relation to its function and aesthetics.
5. Describe a methodology test that can be employed when evaluating the functional criteria for the following:
   - a fishing rod.
   - an energy drink.
   - a toothbrush.
   - a remote control device.
   - socks.
6. Assess the impact of a chosen child's toy on society and the environment.

Sample Preliminary questions

Extended-response question A p. 206

1. Below is a drawing of a child's toy designed as a teaching aid for personal development. Unfortunately the detailed nature of the shape has made it a choking hazard for infants and the product was recalled. Analyse why designers recognise the importance of social responsibility when designing products, systems and environments. (15 marks)

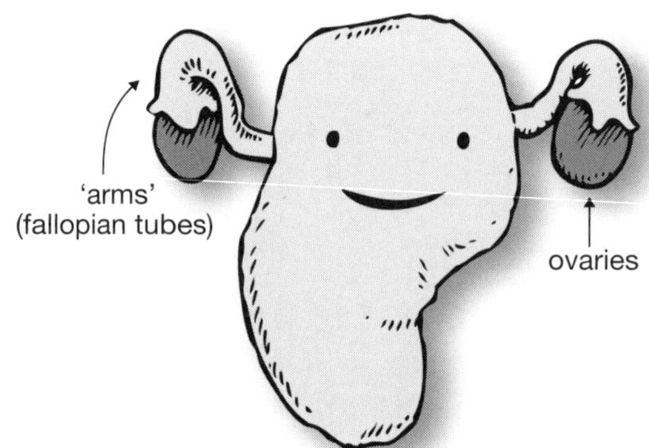

Small part choking hazard
The ovaries may be pulled off and become a choking hazard.
Keep away from children.

10 – Communication

PRELIMINARY OUTCOMES

A student:	You learn about:	You learn to:
P5.2 communicates ideas and solutions using a range of techniques.	■ communication – forms of communication including verbal, written, graphical, visual, audio – elements of the communication process which include sender, receiver, medium, message – criteria for evaluating communication including clarity of message, appropriateness of method chosen and ease of interpretation – communicating information through a variety of media – visualising solutions – the purpose of prototypes and/or models – presentation techniques suited to the needs of design clients and design projects.	■ use appropriate design and technology terminology. ■ experiment with a range of techniques and forms to visualise and communicate ideas and solutions. ■ communicate design ideas and solutions effectively using a range of technologies. ■ use appropriate standards and conventions to visualise and communicate ideas and solutions. ■ justify the selection and use of communication techniques.
P6.2 evaluates and uses computer-based technologies in designing and producing.	■ computer-based technologies and their application including – modelling – research – simulation and graphics – communication – presentation.	■ discriminate in the choice and use of computer-based technologies to develop, communicate and present design ideas and processes.

10.1 Communication

Communication is defined as 'the interchange of information by speech, writing or signs'. It is a process of sending messages from one entity to another. These interactions are between at least two beings which share a collection of symbols and semiotic rules. Although there is such a thing as one-way communication, this occurs when the receiver cannot interpret the meaning of the message. In most instances communication is two-way.

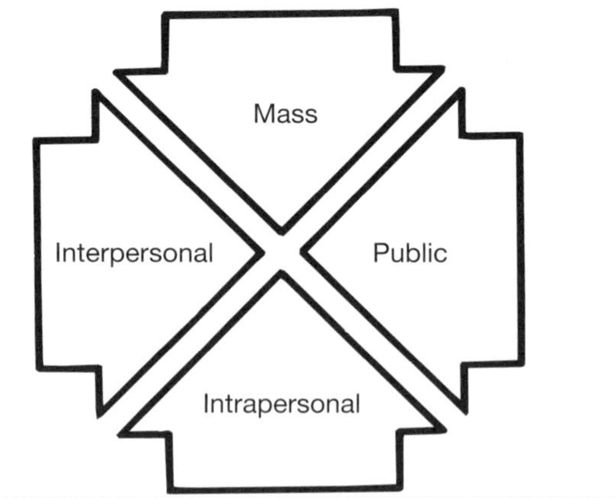

Figure 10.1 Levels of communication

10.1.1 Forms of communication including verbal, written, graphical, visual and audio

People communicate with each other to send a message in a mutually acceptable context. As a result there are a variety of forms and types of communication. The American expert in communicology, John Dwyer, categorises these into three forms of communication—verbal, nonverbal and graphic—and four types of communication: intrapersonal, interpersonal, public and mass communication.

Verbal is oral communication involving speech. Verbal communications may be informal or formal and may be presented by one individual to another or by the use of technology such as video, podcast, Skype or webcasts. Formal examples of verbal communication occur when speeches are prepared and spoken and when thoughts are presented during meetings and interviews. An informal example may be when two people are simply chatting about a topic of choice.

Nonverbal communication refers to body language, signs and gestures, eye contact and facial expressions. Graphic communication is about presenting research and data using images, sketches, photographs, tables and charts.

Intrapersonal communication is when you sketch your ideas on paper to see what they will look like. Interpersonal is communication between people involved in the communication. Public communication may involve making an announcement, distributing flyers, putting up posters or writing letters and emails. Mass communication is used to communicate with large numbers of people. This is usually done using the media.

People usually have a preference for a particular communication style. When these styles are identified effective and predictive communication occurs. Common styles include visual style, where you communicate best through seeing and showing; auditory style, where you communicate best through talking and listening; kinaesthetic style, where you communicate best through feelings and doing; and self talk style, where you communicate best through internal thinking. We all use each style but most people display a preference for one over the others.

10.1.2 Elements of the communication process which include sender, receiver, medium and message

Communication is a process where an information parcel (called the message) is sent. The type of material used to send the message, for example, the telephone, a letter, a poster or picture, is commonly called the medium. The person who determines what the message will be and starts initial communication is called the sender and the person who listens to the message is called the receiver. When they respond, they become the sender and the first sender is now the receiver. This is a continuous process during communication and it involves developing a shared understanding.

Communication requires using a collection of processing, listening, observing, speaking, questioning, analysing and evaluating skills. It is through clear communication that humans can work together and build understandings.

A message or information is sent from a sender to a destination. The sender's and receiver's personal understandings and experiences may cause the message to be interpreted differently depending on their traditions, cultures or gender, which may alter the intended meaning of the message's content.

One problem with this encode-transmit-receive-decode model is that the processes of encoding and decoding imply that the sender and receiver each possess similar understandings and this may create conceptual difficulties. There are common barriers to successful communication, two of which are message overload (when a person receives too many messages at the same time), and message complexity.

Factors impacting on human communication

There are three factors in face-to-face human communication: body language, voice tonality and words. According to research, fifty-five per cent of impact is determined by body language (postures, gestures and eye contact), thirty-eight per cent by the tone of voice, and seven per cent by the content or the words. People recall the way the message was delivered more than the topic itself.

10.1.3 Criteria for evaluating communication including clarity of message, appropriateness of method chosen and ease of interpretation

Good communication skills require a high level of self-awareness. Understanding your personal style of communicating will go a long way towards creating lasting impressions on others. By becoming more aware of how others perceive you, you can adapt more readily to their styles of communicating. You can make another person more comfortable with you by selecting and emphasising certain behaviours that fit within societal expectations. There are four basic communication styles:

- aggressive
- passive
- passive/aggressive
- assertive.

It is important to understand how your communication style is interpreted by others to avoid miscommunication and misunderstandings. The goal is to communicate with assertion but avoid an aggressive, passive-aggressive or passive style of communication.

Aggressive communication

This is when you choose and make decisions for others. You are brutally honest, direct and forceful, self-enhancing and derogatory. You feel righteous, superior and controlling, though later you possibly feel guilt. Others feel humiliated, defensive, resentful and hurt around you; they view you in the exchange as angry, vengeful, distrustful and fearful. The outcome is usually that your goal is achieved at the expense of others. Your rights are upheld but those of others are violated and your underlying belief system is that you have to put others down to protect yourself.

Passive communication

This allows others to choose and make decisions for you. You are emotionally dishonest, indirect, self-denying and inhibited. If you get your own way, it is by chance. You feel anxious, ignored, helpless, manipulated, angry at yourself and/or others, and others feel guilty or superior and frustrated with you. They view you in the exchange as a pushover and that you do not know what you want or how you stand on an issue. The outcome is that others achieve their goals at your expense. Your underlying belief is that you should never make someone uncomfortable or displeased.

Passive-aggressive communication

This allows you to manipulate others to get your way. You appear honest but underlying comments confuse others and you tend towards indirectness with the air of being direct. You are self-enhancing but not straightforward about it and in win-lose situations you will make the opponent look bad or manipulate it so you win. If you do not get your way you are likely to make snide comments or pout and play the 'victim'. You feel confused but also unclear on how you feel; you are angry but not sure why. Later you possibly feel guilty. Others feel confused, frustrated, not sure who you are or what you stand for or what to expect next. Others view you in the exchange as someone they need to protect themselves from as they fear being manipulated and controlled. The outcome is that the goal is avoided or ignored as it causes such confusion or the outcome is the same as with an aggressive or passive style. Your underlying belief is that you need to fight to be heard and respected. If that means you need to manipulate or be passive or aggressive, so be it.

Assertive communication

This allows you to choose and make decisions for yourself. You are sensitive and caring, honest, self-respecting, self-expressive and straightforward. You convert win-lose situations to win-win ones and are willing to compromise and negotiate. You feel confident, self-respecting, goal-oriented, valued, and later you may feel a sense of accomplishment. Others feel valued and respected and view you with respect, trust and understand your positions. Outcomes are determined by above-board negotiation. Your underlying belief is that you have a responsibility to protect your own rights. You respect others but not necessarily their behaviour.

People behave and respond depending on the situation. When communication is difficult it can be helpful to take an approach that suits other people's styles and habits. Once people are aware of the areas needing improvement they can develop new skills to increase the flexibility of their behavioural range and enhance the quality of their communications.

10.1.4 Communicating information through a variety of media

Media is specifically designed to reach large groups of people. Mass media now also includes communications carried out on the Internet, such as blogs, message boards, podcasts and video sharing, because individuals now have a means to exposure that is comparable in scale to that previously restricted to a select group of mass media producers.

Media can be used for purposes such as promoting business and social issues. This can include marketing and advertising, entrepreneurialism, entertainment and Internet sites such as YouTube and Facebook.

Electronic media and print media include radio and television broadcasting, the Internet, blogs, podcasts and mobile phones, which are now referred to as the '7th mass media' because they are used for rapid breaking news, short clips of entertainment such as jokes, and for news alerts, horoscopes, games, music and advertising. Electronic publishing has also developed into a mass form of media with devices such as PlayStation and the Nintendo Wii broadening their use.

10.1.5 Visualising solutions

Communication within design processes involves the visualisation of clear design ideas and solutions. This communication must be clear as it is a shared process and if the solution is not commonly understood the entire process can be a disaster.

During the design process the client will initially put forward some initial ideas. If these are suggested verbally the designer will need to sketch these concepts to ensure that everyone has a common starting point with no misunderstandings. A visual graphic which can be adapted and enhanced is important because listeners often interpret a described idea differently to the speaker's intentions.

The first stage in the design process involves gathering initial ideas, which may mean looking for

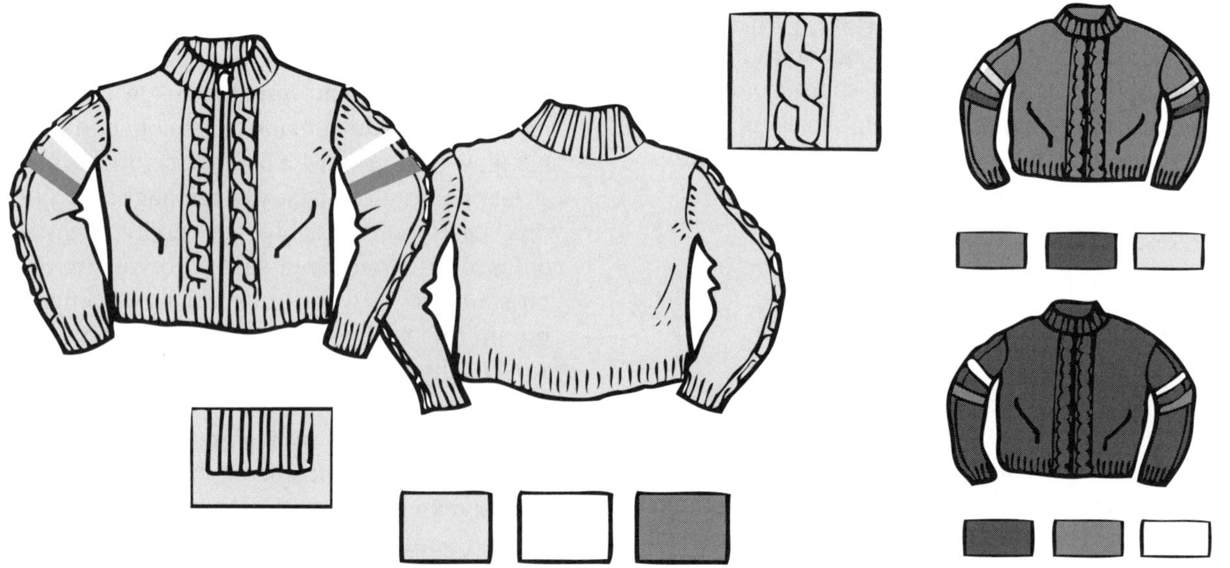

Figure 10.2 Production sketch of a fashion item

possible solutions on the Internet. This is the starting point. It is necessary to include a range of ideas so the designer does not focus on one idea alone and ignore other options.

After research, testing and experimentation is carried out on materials, tools and techniques the original sketches and ideas are examined, then adapted as the final design idea is reached. As ideas merge and the latest technological advances are inserted into the design a final production sketch is produced.

The production sketch includes a quality, coloured front and back view, fully labelled with details and fashion features.

An exploded view is often used to allow the client to visualise how the design will fit together. This sketch also clarifies for the manufacturer how the pieces will need to be manufactured. Sketches are drawn from a range of angles to limit errors.

Figure 10.3 An exploded view of a razor demonstrating how layers lock together.

The importance of clear sketching to visualise solutions both in industry and in your HSC design folio cannot be underestimated. It is this clear sketching that will allow the HSC marker to see that you have considered every option as a design solution and explored the better choices, coming up with an innovative and improved design solution that uses the best ideas, materials, tools, techniques and technologies.

10.1.6 The purpose of prototypes and/or models

A prototype or model is often used as part of the product design process to allow engineers and designers the ability to explore alternatives, test theories and confirm performance prior to starting production of a new product. An interactive series of prototypes will be designed, constructed and tested as the final design emerges and is prepared for production. A common strategy is to design → test → evaluate → modify the design based on analysis of the prototype.

It is common to label the prototype using Greek letters. For example, a first prototype may be called an Alpha prototype. Subsequent prototyping names such as Beta, Gamma, and so on will be expected to resolve issues and perform closer to the final production. There are four types of prototypes or models as described below.

Proof-of-principle prototypes

This type is also called a breadboard. It is used to test a part of the design without copying the appearance, materials or manufacturing process. Proof-of-principle prototypes are used to identify which designs require further development and testing.

Form study prototypes

This prototype allows designers to explore the aesthetics of a product without copying the function. They can assess ergonomic factors. Form study models are easily sculpted and use inexpensive materials, without replicating the final colour, finish or texture.

Visual prototypes

These capture the intended design aesthetic and simulate the appearance, colour and surface textures of the intended product, but they are not functional. They are suitable for use in market research as packaging mock-ups and for promotional photo shoots.

Functional or working prototypes

These simulate the final design in aesthetics, materials and function but are reduced in size to reduce costs. The construction of a working full-scale prototype is the engineers' final check for flaws and it allows last-minute improvements to be made before larger production runs are commenced.

With the recent advances in computer modelling it is becoming practical to eliminate the creation of a physical prototype by using a computer model instead. Computer modelling is now being extensively used in automotive design, both for form (in the styling and aerodynamics) and for function.

10.1.7 Presentation techniques suited to the needs of design clients and design projects

When presenting ideas and the final solution to the client the presentation will contribute to the success or failure of the design. Some hints for successful presentations are listed below.

- Use visual aids. The use of graphics, video clips, appropriate humour and images in presentations increases interest, focuses on the design solution, and improves the chance of meeting the objectives.
- Keep it short and sweet. An old adage says 'No-one ever complained of a presentation being too short'. Nothing kills a presentation more than going on too long. Keep oral presentations to less than twenty minutes.
- Use the 'rule of three'. People tend to only remember three things. Work out what are the three messages you most want your audience to grasp and structure your presentation around them. In visual terms, use a maximum of three points on each slide during presentations.
- Practice makes for perfect performance. Rehearsal is the biggest single thing that you can do to improve your performance. Perform your presentation out loud at least four times, once in front of an audience.
- All presentations are a type of theatre. Use narratives and tell stories and anecdotes to help illustrate points.
- Bullet points are the kiss of death for most presentations. Most people use bullet points as a form of speaker notes. To make your presentation more effective put your speaker notes in your notes and not up on the screen.
- Set up a video camera and video yourself presenting. You will see the mistakes you are making and will be able to correct them before the presentation.
- You should always know what slide is coming next. It sounds very powerful when you say 'On the next slide [Click] you will see …' and that is what appears.
- Have a back-up plan. Murphy's Law normally applies during a presentation with technology not working, power cuts, the projector blowing a bulb, spilling coffee on your shirt, not enough power leads, no loudspeakers, presentations displaying strangely on the laptop, and so on. So, take along the following items: a printed out set of slides, a CD or data stick of your presentation, and a laptop with your slides on it.
- Evaluate the presentation room. Arrive early and make sure you see your slides working on the screen. Work out where you will need to stand.

10.2 Computer-based technologies and their application

The computer, the size and weight of a small, thin book, incorporates television, radio, music and mobile phone technologies, along with wireless Internet access for email, texting, information gathering and instant messaging.

Computer-based technology applications include modelling, research, simulation and graphics, and communication and presentation skills. All of these applications are commonly used during the design process.

10.2.1 Modelling

Computer modelling or simulation is a computer program that promotes the imitation of a real product or process. The act of simulating something involves copying the key characteristics and behaviours of a selected physical or abstract product, system or environment. Computers allow modelling to occur in three dimensions and allow movement within the model.

Computer modelling and simulations are used for mathematical modelling of many natural systems in physics, astrophysics, chemistry and biology, as well as human systems in economics, psychology, social science and engineering. Simulations can be used to explore and gain insights into new technology and to estimate the performance of systems too complex for analytical solutions.

Other contexts include simulation of technology for performance optimisation, safety engineering, testing, training and education. Simulation can be used to show the eventual real effects of alternative conditions and courses of action. Simulation is also used when the real system cannot be engaged because it is inaccessible, may be dangerous or unacceptable to engage, or it may simply not exist.

Within your MDP, computer modelling is useful for showing the HSC markers a working 3-D model of your design solution. This computer simulation can be used as the actual project if manufacturing is impractical, or as a tool to demonstrate the internal actions of the project.

10.2.2 Research

Research can be defined as the search for knowledge or any systematic investigation to establish facts. The primary purpose of research is discovering and interpreting as well as the development of methods and systems for the advancement of human knowledge.

The goal of the research process is to produce new knowledge. There are three main types of research. These are:

- exploratory research, which structures and identifies new problems
- constructive research, which develops solutions to a problem
- empirical research, which tests the feasibility of a solution using empirical evidence.

For the MDP both primary and secondary research is expected to be carried out.

10.2.3 Simulation and graphics

Computer graphics

For computer simulation refer to computer modelling above.

The term 'computer graphics' has been used to describe almost everything on computers that is not text or sound and refers to:

- the representation and manipulation of image data by a computer
- the various technologies used to create and manipulate images
- the images produced
- the sub-field of computer science, which studies methods for digitally synthesising and manipulating visual content.

The development of computer graphics has allowed a better understanding and interpretation of data. Developments in computer graphics have revolutionised animation, movies and the video game industry.

Computer graphics touch many aspects of daily life: on television, in weather reports in newspapers, and in many types of medical investigations and surgical procedures. A well-constructed graphic can present complex statistics in a form that is easier to understand and interpret.

Computer-generated imagery

Computer-generated imagery can be categorised as 2-D, 3-D, 5-D and animated graphics. As technology has improved, 3-D computer graphics have become more common, but 2-D computer graphics are still widely used for the computer-based generation of digital images. Two-dimensional computer graphics are mainly used in applications that were originally developed from traditional printing and drawing technologies. Two-dimensional models are often preferred because they give more direct control of the image than 3-D computer graphics, whose approach more resembles photography.

Pixel art is a form of computer graphic. It is created through the use of raster graphics software, where images are edited on the pixel level. Graphics in most old computer and video games and many mobile phone games are mostly pixel art. In contrast, 3-D computer graphics use a 3-D representation of geometric data that is stored in the computer for the purposes of performing calculations and rendering 2-D images.

Computer animation is the art of creating moving images via the use of computers. It is created by means of 3-D computer graphics, though 2-D computer graphics are still widely used. To create the illusion of movement, an image is displayed on the computer screen then quickly replaced by a new image that is similar to the previous image but shifted slightly. This technique is identical to the illusion of movement in television and mot images.

Graphics are visual presentations on a surface such as a wall, canvas, computer screen or paper that are used to brand, inform, illustrate or entertain. Examples include photographs, drawings, line art, graphs, diagrams, typography, numbers, symbols, geometric designs, maps, engineering drawings, and so on. Graphics often combine text, illustrations and colour. Graphic design may consist of the deliberate selection, creation or arrangement of typography alone, as in a brochure, flier, poster, website or book, without any other element.

Two and 3-D scene files contain objects in a strictly defined language or data structure. The files contain geometry, viewpoint, texture, lighting and shading information as a 'description' of the virtual scene. The data contained in the file is then passed to a rendering program to be processed and output. The rendering program is usually built into the computer graphics software.

10.2.4 Communication

For additional information on communication refer to pages 96–99 of this chapter.

10.2.5 Presentation

For additional information on presentation techniques refer to page 100 of this chapter.

Key concepts and definitions

Communication—sometimes called 'communicology', it relates to all the ways we communicate. Communication includes both verbal and nonverbal messages.

Graphical communication—image-based communication such as digital sketches and images, video clips, scans, tables, charts and pictures.

Prototypes—models, often used as part of the product design process to allow designers to explore alternatives, test theories and confirm performance prior to starting production of a new product.

Useful websites

- laconicdesign.net/featured/the-power-of-sketching
- www.cce.usyd.edu.au/binary?id=731942
- www.drbackman.com/communication-styles.htm
- www.dwconline.com/DWC/August'00/designs.html
- www.nhaustralia.com.au/search.aspx?keyword=Business+ Communication&gclid=CMqhloe6gqACFSOjagod9zbfCA

Classroom activities

Sketch the following:

1. Thumbnail sketches of a range of theme park rides.
2. A concept sketch of an irregularly-shaped building.
3. An exploded view of an engine or a hamburger.
4. A production sketch of a formal dress, a television cabinet, a wedding cake or a laptop holder.

Key concept questions A p. 206

1. Differentiate between forms and types of communication.
2. Describe the purpose of a prototype or model.
3. Explain how clear communication is essential to the success of a design project.

Sample Preliminary questions

Objective-response questions A p. 207

Circle A, B, C or D. (1 mark each)

1. Clear verbal communication involves
   A oral communication that may be formal and/or informal.
   B sending a mixed message through the medium to an encoder who will decode it.
   C nonverbal body language, signs and gestures, eye contact and facial expressions.
   D graphic communication.

2. Select the statement which is correct. Common communication styles may include
   - A visual style where you communicate best through seeing and showing.
   - B auditory style where you communicate best through talking and listening.
   - C kinaesthetic style where you communicate best through a presentation.
   - D a combination of the above with a focus on verbal communication.
3. The elements of the communication process include
   - A interpersonal, intrapersonal, public and mass communication.
   - B sender, receiver, medium, message.
   - C sender, receiver, communication, message.
   - D interpersonal, intrapersonal, public and mass communication, sender, receiver, medium, message and sender, receiver, message.
4. The major factor in promoting clear and memorable communication is
   - A tone of voice.
   - B content of the message.
   - C body language.
   - D a clear message.
5. When presenting design solutions to a client it is important to use the rule of three because people only remember three things. This will involve
   - A only speaking to three people.
   - B using three main points on a slide that you want people to remember.
   - C repeating your three points three times in the presentation.
   - D repeating the same message for three days in a row.

Short-answer questions A p. 207

6. Clear communication is essential to working in a collaborative design team. Explain what is meant by the terms 'collaborative' and 'communication'. (4 marks)
7. List the two main barriers to successful communication and explain why they are barriers. (4 marks)

Extended-response questions A p. 207

8. When presenting ideas and the final solution to the client the presentation can add to the design's success or failure. Suggest ways in which a successful presentation may be ensured. (15 marks)
9. Consider the following situations and outline how:
   - a passive person would react
   - an aggressive person would react
   - a passive-aggressive person would react
   - an assertive person would react.

   (a) You are trying to concentrate on some important work. However, a few of your co-workers are laughing and horsing around. What do you do? (7 marks)

   (b) Your friend has borrowed your laptop to do some work. He has had it for several hours and it is now time to go home. You really want to take it home to do some folio work. What do you do? (8 marks)

HSC course

11 – Factors affecting the success of design projects

HSC OUTCOMES

A student:	You learn about:	You learn to:
H1.1 critically analyses the factors affecting design and the development and success of design projects.	■ factors affecting designing and producing, including – appropriateness of the design solution – needs – function – aesthetics – finance – ergonomics – occupational health and safety – quality – short- and long-term environmental consequences – obsolescence – life cycle analysis.	■ apply factors affecting design to the development of the major design project. ■ debate the issues and factors influencing design and design practice.
	■ examples of success and failure in design.	■ critically analyse examples of success and failure in design solutions.

The information provided to support the HSC Outcomes build on what you have learned in the Preliminary course. It is prudent, at this point, for you to revise Chapter 2, which provides the starting point for factors affecting success. This chapter moves into factors affecting designing and producing in commercial settings. Use this information and apply it to your major design project.

11.1 Factors affecting designing and producing

In commercial settings design is affected by organisational, social, cultural, environmental and behavioural factors. Organisational factors include the characteristics of the task, work flow, ergonomics and manufacturing and work practices. Social and cultural factors include personal values, cultural beliefs, and individual and community needs. Environmental factors include aspects related to the natural world, a responsible ethic in relation to material use, a biocentric ethic, sustainable development, eco-design and life cycle analysis. Behavioural factors include ethical and responsible design, feedback, autonomy, use of abilities, the use of a variety of sustainable resources and ongoing evaluation.

11.1.1 Appropriateness of the design solution

The appropriateness of the design solution must be thoroughly investigated to ensure its innovative nature, its degree of difference from existing products, and for quality in design and manufacture. The environmental impacts of designs are being increasingly evaluated in order to ensure responsible design. This can be achieved through ongoing evaluation at each stage of the project's evolution. The major design solution must be managed throughout the entire design process to ensure an innovative quality product.

In the mind map below 'managing the innovation' is the central focus with each factor that impacts on its success rotating around it. These factors impact on each other as well as the innovation in the centre.

11.1.2 Needs

When starting a design project the need or problem must be thoroughly researched and explored before possible solutions can be found. The need must be broken down into its components and every element examined.

11.1.3 Function

The first criteria that is used to evaluate success is that of function. A design solution can only be successful if it does what it is meant to or it solved the initial problem. If it does what it is meant to it is said to be functional.

11.1.4 Aesthetics

The aesthetics of the design refers to the properties of the project related to appearance. The appearance can be described in terms of the elements of design, such as shape, size, line, direction, colour and texture.

11.1.5 Finance

Whenever a new product, system or environment is being created there is a need to plan and manage a budget. The budget must include income and

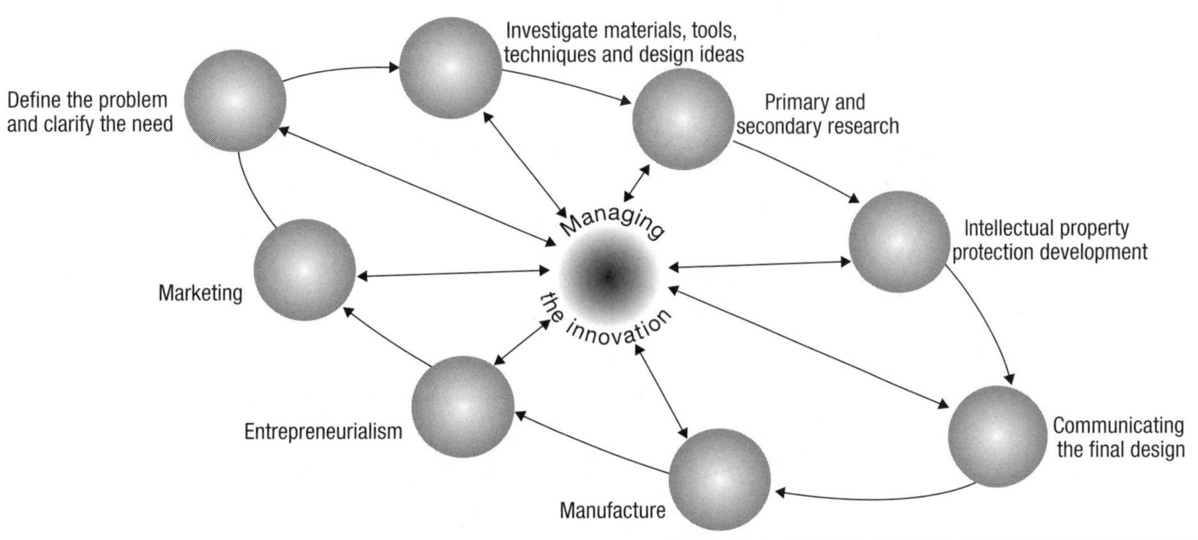

Figure 11.1 Managing the design process

expenditure for all items. The income must include a list of all financial sources including donations and sponsorships. Expenditure includes the cost of all items such as tools, materials, the manufacturing processes, labour, energy and time.

11.1.6 Ergonomics

The ergonomics of a design refers to how it is fitted to the structure of the body. When creating an ergonomically-designed item the designer must consider the anthropometrical measurements. These measurements consider the length and size of every part of the body. The designed item then supports the physical structure of the human skeleton and muscles to ensure the body is held in a safe supported position.

11.1.7 Occupational health and safety

For information on this topic please refer to Chapters 2, 8 and 16.

11.1.8 Quality

The quality of the product will have a direct consequence on whether it is successful. A quality product will have good design that achieves what is expected and it will be manufactured from environmentally sustainable materials obtained from a sustainable source. In addition, the manufacturing process will have a minimal impact on the environment and the product will have been evaluated thoroughly for both its functional and aesthetic considerations.

11.1.9 Short- and long-term environmental consequences

Sustainable technologies are those which use less energy, fewer limited resources, do not deplete natural resources, do not directly or indirectly pollute the environment, and can be reused or recycled at the end of their useful life. There is a significant overlap with appropriate technology, which emphasises the suitability of technology to the context, in particular considering the needs of people in developing countries. However, the most appropriate technology may not be the most sustainable one, and a sustainable technology may have high cost or maintenance requirements that make it unsuitable as an 'appropriate technology'. If designers are to become ethically and environmentally responsible then they must become sustainable designers. Some common principles for sustainable design are as follows.

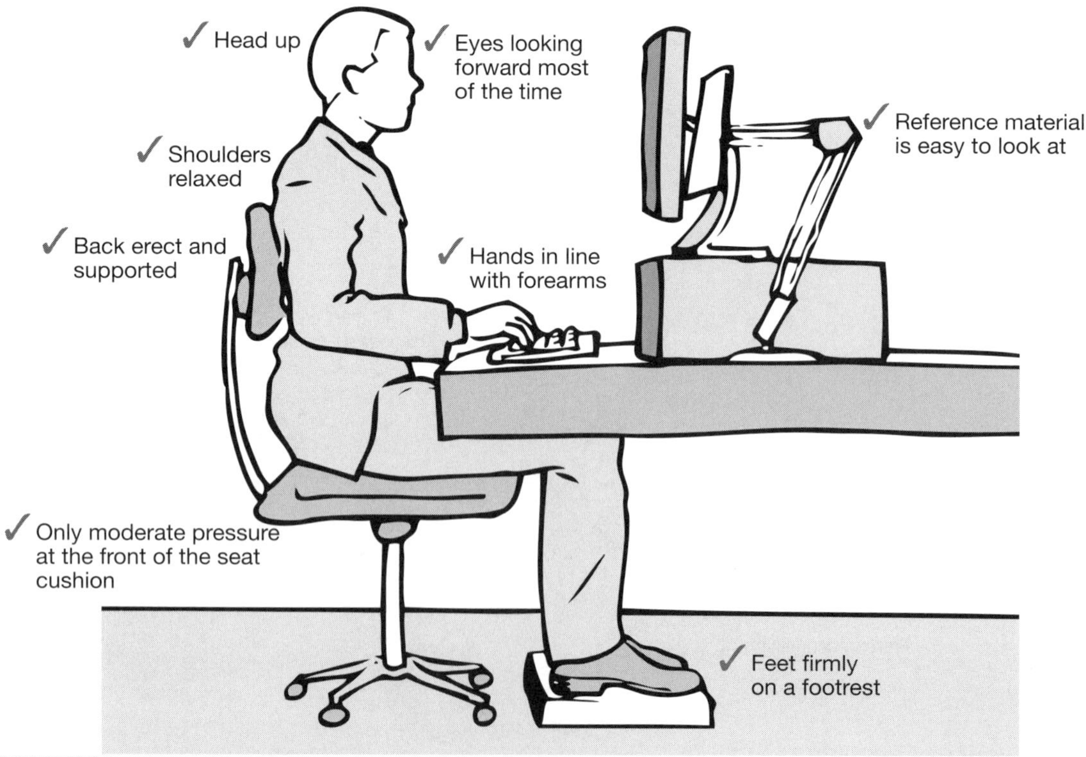

Figure 11.2 Ergonomic desk using anthropometrical measurements

- Low-impact materials: choose non-toxic, sustainably produced or recycled materials which require little energy to process.
- Energy efficiency: use manufacturing processes and produce products which require less energy.
- Quality and durability: longer-lasting and better-functioning products will have to be replaced less frequently, reducing the impacts of producing replacements.
- Design for reuse and recycling: products, processes and systems should be designed for performance in a commercial 'afterlife'.
- Impact measures for total earth footprint and life cycle assessments for all resource uses are increasingly available. Many are complex, but some give quick and accurate whole earth estimates of impacts.
- Renewability: materials should come from nearby (local or bioregional) sustainably managed renewable sources that can be composted (or fed to livestock) when their usefulness has been exhausted.
- Healthy buildings: sustainable building design aims to create buildings that are not harmful to their occupants or the environment. An important emphasis is on indoor environmental quality, especially indoor air quality.

11.1.9 Obsolescence

Obsolescence occurs when a person, object or service is no longer wanted even though it may still be in good working order. Obsolescence frequently occurs because a replacement has become available that is superior in one or more aspects.

Technical obsolescence is when a new product or technology supersedes and becomes preferred to the old item. Items may become functionally obsolete when they do not perform in the manner that they did when created. Products which naturally wear out or break down may become obsolete if replacement parts are no longer available, or when the cost of repairs or replacement parts is higher than the cost of a new item.

Sometimes marketers deliberately plan obsolescence into their product strategy with the objective of generating long-term sales volume by reducing the time between repeat purchases. For more information on this topic please refer to Chapter 2.

11.1.10 Life cycle analysis

Life cycle analysis or assessment includes the entire life cycle of the product, encompassing extracting and processing raw materials; manufacturing, transportation and distribution; use, reuse and maintenance; and recycling and final disposal.

Further information on this topic can be found in Chapter 5.

11.2 Examples of success in design

As computer technology evolves and improves we can see the transformation of conceptual boundaries between traditional design disciplines. As people's needs evolve so will the type of computer that they desire. These can be PCs and laptops with better and faster functions and personal communication devices that include features such as laser-activated keyboards.

The next step in computer technology evolution is the integration of human movement with technology incorporated into game consoles. This is commonly seen in Wii and PlayStation gaming consoles. Virtual reality is concerned with the design, evaluation and implementation of interactive computing systems for human use, and with the study of the phenomena surrounding them. Virtual reality is also used to analyse movement, with the aim of improving sporting actions and achievements. For example, stroke correction in swimming, golf and tennis.

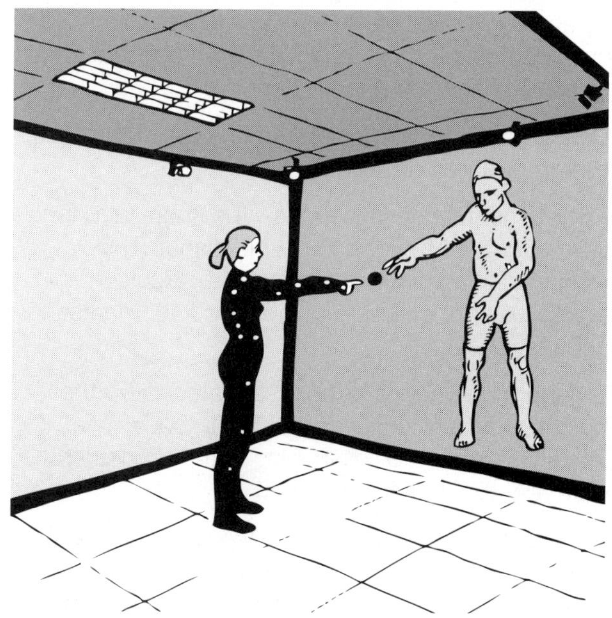

Figure 11.3 Virtual reality

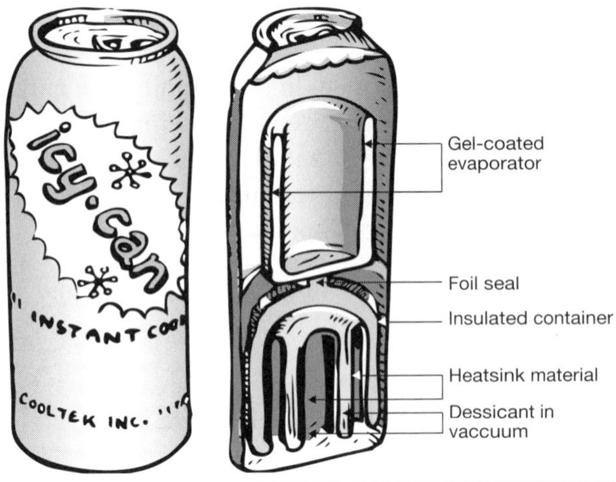

Figure 11.4 Self-chilling can

Figure 11.5 Clothing labels for the vision impaired

Key concepts and definitions

Appropriate technology—technology which, in its creation and use, meets human needs while considering the short- and long-term consequences for society and the environment.

Design solution—the created answer to the design problem or need.

Environmental consequences—the short- and long-term effects of the design on the planet. This includes all factors such as material selection, transportation, processing, by-products, marketing and distribution.

Ergonomics—how a design is created to be the best fit, size and shape for a human body. The design is created by examining anthropometrical data and the measurements of the body.

Successful design means that the plan, proposal or invention has achieved its goal in solving a problem or meeting a design brief. Success can be judged financially or in terms of problem-solving or environmental responsibility.

Responsible design demonstrates a commitment to environmental sustainability through product life cycle analysis, a strong environmental understanding, a focus on ethics related to material use, a biological focused ethic, sustainable development, eco-design and life cycle analysis.

Useful websites

- www.baddesigns.com/examples.html
- www.cleardesignuk.com/design-brief.html
- www.lisi.usb.ve/publicaciones/10%20 proceso%20de%20desarrollo/proceso_03.pdf

Classroom activities

1. Describe how each design factor has been considered in your MDP.
2. Discuss how each design factor has either contributed to or hindered the success of your MDP.
3. If you were to improve and redesign your MDP:
   - suggest which emerging technologies, materials and design ideas could be included.
   - research alternate manufacturing techniques and evaluate their worth.
   - discuss how sustainable and responsible design would further impact on this solution.

Key concept questions A pp. 207–208

1. List the factors that determine whether a design will be successful or fail.
2. Describe the different types of obsolescence.
3. Discuss the organisational, environmental, social and cultural factors affecting designing and producing.
4. Explain the link between a product's life cycle and becoming a responsible designer.

Sample HSC questions

Extended-response questions A p. 208

1. The 'tag' is a soft-shell mobile phone. It is a malleable and 'casual' communication device. It is not only soft but also flexible and can be hung from a belt or wrapped around the user's arm. Shape-memorising materials and multiple pressure sensors allow the phone to change its shape according to requirements.

   Discuss the factors to be considered that will contribute to either the success or failure of this design.
   (15 marks)

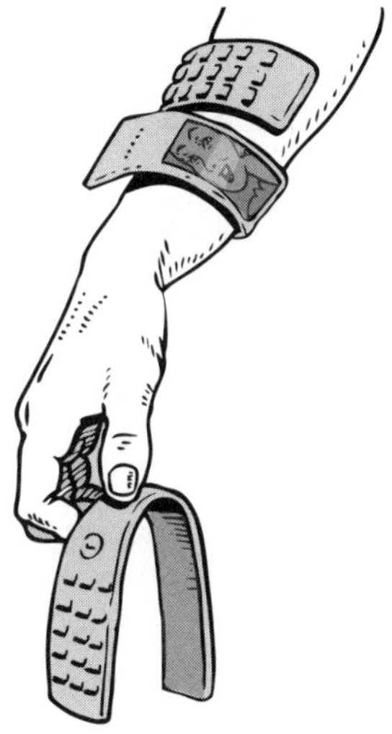

2. For your MDP critically analyse, in terms of responsible design, each of the factors affecting the design, development and success of the design project. (15 marks)

12 – Practices and processes of designers

HSC OUTCOMES

A student:	You learn about:	You learn to:
H1.2 relates the practices and processes of designers and producers to the major design project.	■ the work of designers – design practice – processes used by designers.	■ emulate, where appropriate, the practices and processes usesd by designers to assist in the development of the major design project.

12.1 The work of designers

12.1.1 Design practice

Products in the current global marketplace are being produced at an accelerating rate. No longer are goods limited to a localised market: the advent of the Internet and its network of consumers allow products to be widely distributed from any virtual space through websites such as eBay and amazon.com. This has provided challenges to designers in keeping up with societal expectations but also provided opportunities to respond to a more inventive and spontaneous culture than the world has ever known in the past. Successful Australian designers such as Sally Dominguez, Toby Grime, Alison Page, Angelo Kotsis and Marc Newson have used vision, creativity and initiative and their achievements in design thinking have developed their global reputations.

Whether you are a global designer or a HSC Design and Technology student developing your MDP all designing goes through similar design practices: it is the design thinking that shows up the differences. Good design requires consideration of the elements and principles of aesthetics and function as well as other factors that will affect the final product. It is a fluid process of thought, research, modelling and adjustment leading to a final design or a redesign. However, designing can still display a random nature because of the possibility of generating ideas.

Like many designers, HSC students see designing as a continual management of constraints. An important step is to recognise these constraints and classify them as 'negotiable' or 'non-negotiable'. By this a designer can manipulate design variables to satisfy non-negotiable constraints while optimising those which are negotiable. In understanding a non-negotiable constraint when designing a chair, the designer must recognise that a chair must support a certain weight to be useful, and this is a non-negotiable constraint. This will impact on the design thinking regarding the solution. Another non-negotiable constraint would be cost. Many Design and Technology students are concerned about the cost of their final design solution and decisions based on cost will have an impact on their final work. While these factors seem limiting to the design it is the choice of contemporary materials and the aesthetic qualities that will remain as variables. Designers and students can consider these as 'negotiable' and explore them further.

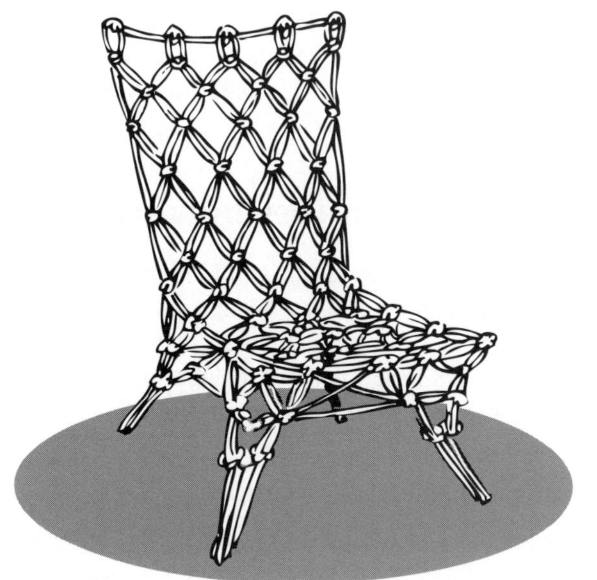

Figure 12.1 Knotted Chair

The Knotted Chair (1996) made by Marcel Wanders was quite radical at the time and won him design and furniture awards. The furniture company Cappellini agreed to mass produce the product. It is innovative in that it is made from reinforced carbon fibre rope. For the rope to hold its rigid form it was dipped in epoxy resin.

Wanders is clever in the way he takes high technology and mixes it with high design and the Knotted Chair is an example of where contemporary materials and the aesthetic qualities will remain as variables.

Poor designs are generally the result of mismanaged constraints. The results tend to be less rewarding for the user in the long term. Cheap and ugly products can become highly visible in the world and do not sit well in the balance of design. Good design recognises each constraint and balances it as part of the overall decision-making process. As with any creative process, a Design and Technology student can design anything! Design is simply the management and choice of constraints; the process of timely decisions during many stages.

Design practice results from the creativity, innovation and problem-solving that forms from the design thinking and is ever-present in most occupations where decision-making or problem-solving is required. This practice is commonly known as the design process. The design process ensures that both the designer and the Design and Technology student are able to work within a sequence of events, providing a strong level of organisation that will eventually lead to the best answer or solution. Similarities between designers include the curiosity, collaboration and end focus that contributes to the overall picture.

12.1.2 Processes used by designers

Strong and flexible design thinking is essential to the designer and also the Design and Technology student. It is important to keep in touch with all aspects of the design without losing the 'big picture'. Some areas of design thinking are influenced by the following:

- attention and focus on customers/users
- redesigning or rethinking the design for the future
- finding alternatives: designing is not about choosing between multiple options; it is about creating those options and bringing new ideas to the problem that others might overlook
- reclaiming design by thinking comprehensively about the entire life cycle of all designed things
- developing the link between design theory and its practical application
- using brainstorming and prototyping to find and test solutions
- 'wicked problems': the problems designers are used to taking on are those without a clear solution, with fuzzy boundaries, and where the outcome is never known and usually unexpected
- design is multi-disciplinary and touches on areas as diverse as psychology, ergonomics, economics, engineering, architecture and art from which to draw inspiration and solutions
- in design, decisions without an emotional component are lifeless and do not connect with people
- bringing intelligent humour into the design when applicable.

Designs are very personal and they provide a connection between the design and person. A well-known designer who recognises the importance of humanist features in designing is Marcel Wanders. The human value is incorporated in his production of families of lamps in Papa, Mama and Baby sizes. Marcel's design process easily allows him to go beyond that of daily life to the place where he can develop his inspirations. In striving for these qualities he places a great deal of significance on the person who purchases one of his products. The end user needs to receive a positive feeling from an object through touch and sight. This feeling develops a bond between the user and the object, one that can remain for a lifetime.

During Marcel's stay in Milan he used his new holiday car, the Antelope II. This extravagant, multicoloured, Bisazza-tiled automobile designed by him in collaboration with Bisazza is a street legal, mechanically-working vehicle covered in a jewel-like Venetian glass mosaic. Keep an eye out next time you are cruising the streets of Milan.

Figure 12.2 Antelope II

Designers such as Marcel Wanders connect design to real people in a focus that is more from the heart and less from the brain. Wanders calls this the 'soul of design' and it is one of his goals in 'making things with love'. In a time of industrial design and high-end technology production many designers worry that mass-produced objects may lose their personal appeal. Designers are careful in maintaining the emotional connection with the object even when working with the most recent techniques and materials; many incorporate something handmade, natural or 'friendly' as a balance. This is an especially important aspect for Design and Technology students as their MDP will certainly reflect their personality and 'touch'.

The long-term relationship that products have with their users is a goal for all to develop in their designs. The concept of an 'aged' product questions designed obsolescence and our throw-away culture of many meaningless products. It is also part of sustainability, ecology and 'green' design. Many designs are designed for the 'now' and support a traditional culture. It is important for the Design and Technology student to not only think of present forms as an inspiration but to recognise the importance of looking to the future. By this, the designs will present as futuristic and enable them to maintain their freshness throughout their life. Many futuristic designs have become outdated but many others are remembered as cult legends. The status of any design will be determined by how it is valued within the surrounding culture.

Many designers employ an avant-garde approach based around a sense of humour. Many students also reflect their personality and humour in their designing. While it is not a necessary element, it will instantly define the product to an audience or user group. Marcel Wanders tries to remind us that at times we must stop, use and reflect and that design is an 'unexpected welcome'. Marcel Wanders' Droog Design is a very successful 'brand' that supports designers who pursue intelligent humour in their designs. It is not just to be laughed at but to bring an emotion to people.

The porcelain egg vase started out as a balloon filled with eggs: that is where the form was taken from. It was then cast and the eggs broken. Marcel Wanders introduces this intelligent humour into his work, and is able to achieve some interesting results to create the 'wow' factor and make people stop and look.

Figure 12.3 The egg vase

Marcel Wanders made the BLO light for Flos, a large Italian lighting firm. It looks like the 'Wee Willie Winkie' candle and when you blow on it you can turn it on and off.

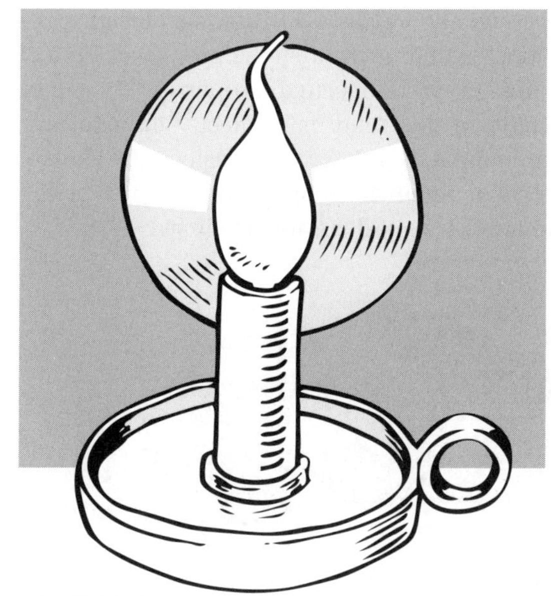

Figure 12.4 The BLO light

Design and Technology students should also recognise the creative value of designing. Creativity (or 'creativeness') is a mental process involving the generation of new ideas or concepts. It can also occur as new associations between existing ideas or concepts. This creative thought is considered to have both originality and appropriateness and makes ideas and concepts fresh. Although intuitively a simple phenomenon, the notion of creativity is in fact quite complex as it is largely about the re-arranging of what we know in order to find out what we do not. It is very important that all creative ideas are pursued through structures such as cognitive organisers, design sketches and concept boards. Some students disregard some of the initial steps and unfortunately their work may suffer further along.

Designing in industry

Designing tends to be seen as personal. This is due to the fact that designing, like decision-making, fulfils an individual purpose or need. There are times where the designing moves beyond individual goals, as seen in companies and industry where design occupies a wider role or for a commissioning agent or client, rather than the designer. Each designer acts according to the design brief set down through company policy, philosophy or mission statement. Different designers will manage the process of the design in many different ways but the core of the process will remain.

It is widely recognised among designers and companies that ideas are to be protected. Protection by means of intellectual property (IP) refers to property of the mind or intellect. Companies go to great lengths to protect their designs. Designers are also held under a loyalty contract for a period of time after design ideas are developed.

Figure 12.5 Protection of ideas is very crucial in design.

The Design Council in the United Kingdom studied the design practices of many leading companies and found some striking similarities and shared approaches. Companies such as Alessi, LEGO, Microsoft, Sony, Starbucks, Virgin, Whirlpool, Xerox and Yahoo are all world leaders in their fields and have a public commitment to the use of design to improve their brand strength and product and service offerings. In design it is imperative to have the right people and processes in place. If Design and Technology students position themselves the same it will assist in responding to management challenges.

Designers in a large industrial setting still maintain strong personal ideas on design. However, they are required to develop their designs for the company in a commercial context. They share an interest in every process that is occurring in the industry and work together with a variety of people across the organisation.

In the past design companies were given an assignment or brief which would be passed on to the designer/s. The result would be revealed from 'under a black sheet' and presented to the client. These days many notable commercial designers have moved some distance from this original notion. Now, all successful design companies are intensively collaborative 'design thinkers'. Collaboration is also important to students as it exposes them to the 'real' world of design and provides an opportunity to share ideas. The real mark of success of a designer, however, is bringing the client alongside as a stakeholder.

Many contemporary designers recognise their responsibilities and realise the danger of becoming sidetracked by the design and not considering pollution and environmental damage. They are also aware of their role in encouraging a consuming lifestyle inspired by marketing. Just as certain environmental practices have changed in contemporary industrial design circles so too should the design practice used by students assess the short- and long-term environmental consequences.

The idea of sustainability is still quite abstract as many of today's initiatives that are called 'sustainable' are, at best, only minimally so in relation to their overall ecological impact. The Eco Design Foundation in Australia recommends to their associates that sustainable contributions are a better goal-driven initiative. 'Sustainments' is a term coined to consider the purpose of sustainable design. It provides clear, assessable indicators. The purpose of sustainable designing is to develop an intuitive sense for sustainability where the entire life cycle of all designed things is considered. It involves the value judgements of designers and students and generally means they will negotiate their design brief with clients to introduce them to the concepts and applications of sustainability. This gives the potential for all designed contributions to become more sustainable. Some of the practical applications of sustainable design theory involve the use of low-impact materials such as non-toxic, sustainably produced or recycled materials which require little energy to process. There is also consideration of the efficient operation of the design once in use.

Designers work through the design process with a trust in their own ability. They calculate the risks and work within constraints. The choices they make are deliberate and can involve ethical choices. Above all they make conscious decisions regarding continual improvement.

The 1930s, 1950s and 1960s saw a strong development in industrial design. While many designers had the luxury of watching their work reach maturity over many years the fast track technology of today means that many designs 'age' in a matter of six months. Consumers and user groups require new designs to be better and to come sooner, meaning that the emphasis for designers is keeping up with the technological demands placed on society.

The relationship between designers and students

The designer	HSC Design and Technology student
Can be employed within an organisation or may be grouped in a collective in a studio or school, or be freelance. Have generally completed tertiary studies or the equivalent in design colleges such as Whitehouse or Billy Blue.	Has had an active interest in design and was able to fulfil this through studying Design and Technology. Has successfully completed two mini design projects in the Preliminary course.
Will be given a design brief by the company or client. The freelance designer might identify a need and commence to solve it.	Throughout the student is supported by the teacher in developing strong ideas about design. The student will identify a need and commence to solve it.
Works through a series of design briefs to identify constraints and classify them as 'negotiable' or 'non-negotiable'.	Works through a series of design ideas identified by a design situation and a brief leading to a proposal. The proposal will clarify ideas and help identify a series of constraints and classify them as 'negotiable' or 'non-negotiable'.
Develops their ideas initially in a sketch book. Generally keeps this book with them as an idea can come along at any time.	Develops their ideas initially in a sketch book and on paper. Collects all loose papers to review and places them into a journal for evidence of planning for their MDP.
Sources existing products by searching websites and patent offices.	Sources for existing products in the world by asking others, investigating stores and searching websites and possibly patent offices.
Attention is on customer needs. Involves designers thinking about redesign, finding alternatives, reclaiming design, wicked problems, sustainable design, and then drawing from a range of influences including inspiration and emotion.	Focusing now on a need. It may not be provided by an external source; rather it may have been developed by a want, a reason or purpose. Involves themselves in design thinking about redesign, finding alternatives, reclaiming design, wicked problems, sustainable design and then drawing from a range of influences including inspiration and emotion.
Meet with clients to clarify ideas and then with experts with knowledge on what is required for the design. Knowledge exchange includes information on material selection, tools required for production and production methods. Experts can provide very sound advice and that often supports that industry.	Gaining inspiration from a variety of areas. Talking to people with knowledge and also experts with or without formal training on the topic.
Develops a strong management plan involving action, time and financial planning. Connect with industry to develop the infrastructure for production of the design. Some designers will follow this process through independently, providing their own financial backing.	Looking at local resources that will be used. Identifying human resources such as the teacher, a parent or interested family member, peers or some experts or mentor. Some experts used as mentors are willing to provide feedback and advice locally or via email. Identifying non-human resources such as rooms and equipment in the school, libraries and the Internet. Connections founded with industry that may be required in case a small part of the production process uses high technology. Students should document such activities. The MDP will require a strongly developed management plan involving action, time and financial planning. Strong understanding of financial needs and commitments.
Return meetings with the client involving cardboard models and appropriate sketching to communicate the idea. Research results will also be supplied as well as any alternative design suggestions. Collaborative meetings with other designers in the same or alternative fields involving modelling and sketching. Talking through the design idea and developing it as 'real' design. Identify materials and associated production methods.	Ongoing meetings with the teacher. Use of cardboard models and appropriate sketching to communicate the idea. Producing alternative design suggestions. Coming up with strategies to produce a final idea. Identify materials and associated production methods. Collaborative meetings are used at all stages as a way of keeping on track. The student could be characterised as an athlete while the mentor is the coach and the teacher the manager.

The designer	HSC Design and Technology student
Agencies may be used to develop the idea from here depending upon the complexity of the technology. Agencies with financial backing can help a designer in the incubation period of the design.	
Final designing distilled down to the development of an accurate working model or prototype. Collaborative meetings discuss the appropriateness of the design and an evaluative process is undertaken. Evaluations may be determined from surveys, working models of a process or replication through software analysis. Final decisions are made and a final meeting with the client completes this process.	Final designing distilled down to the development of an accurate model or working prototype. Collaborative meetings discuss the appropriateness of the design and an evaluative process is considered. Evaluations may be determined from surveys or working models. Final decisions are made.
Tool-up commences and production is confirmed. First batch is produced. The commercialisation is complete with the development of stock inventory, warehousing and distribution of the product. Alternatively, a studio produces an object for sale as a design piece. The final design exists as a product, system or environment and the process is complete.	Production of the MDP commences. All of the design evolution is documented and communicated within the folio. Any variations to the time management or the finance plans are carefully evaluated. The ongoing evaluation will be connected at the end by final evaluations based on a series of parameters. The success of the final design is weighed up against the initial established criteria for success. The final design exists as a product, system or environment and the process is complete.

Key concepts and definitions

Avant-garde—the application of new concepts and techniques in a given field. It can comprise of anything creative, radically new or original in the area of innovation.

Client—a customer who orders or purchases something from someone else. In this case a client asks a designer to design and produce for them.

Constraint—a restriction on the degree of freedom you have in providing a solution. Constraints can be economic, political, technical or environmental and relate to project resources or schedules.

Contemporary—belonging to or being representative of the present time.

Intellectual property (IP)—the property of your mind or intellect. It can be an invention, trademark, original design or the practical application of a good idea. Under law, intellectual property can be protected.

Non-renewable resources—those that cannot be produced, re-grown, regenerated or reused on a scale which can sustain their consumption rate. These resources often exist in a fixed amount, or are consumed much faster than nature can recreate them, such as fossil fuels.

Recycling—processing used materials into new products to prevent waste of potentially useful materials and to reduce pollution.

Sustainability—the strategic management of resources that aims to meet human needs while preserving the environment for the needs of future generations.

Useful websites

- www.ag.gov.au/www/agd/agd.nsf/Page/Intellectual_property
- www.ideo.com
- www.marcelwanders.com
- www.ted.com

Classroom activities

1. Designing could be seen as a series of opportunities and constraints. Draw up a T-chart and list all the opportunities and constraints in your latest design project. Break-down the design constraints even further by identifying what is negotiable and what is non-negotiable.
2. List as many design icons as you can think of and identify the designer. For example, the embryo chair and Lockheed Lounge, designed by Marc Newson.
3. Use a table to categorise the varying materials and processes associated with each design icon.

4. In a group, brainstorm some of the 'ugly' designs associated with a lack of design thought.
5. List ten designer products that you could use in a 'typical' day.
6. 'Google' images for designer products that would draw on an emotional experience.
7. Develop a concept board or cut out a photo-collage to represent unsustainable design.

Key concept questions A pp. 208–209

1. All designers seem to have their own processes and inspirations for creating a new design. Choose a previous design project that you have completed and trace through the processes that brought you through to its successful conclusion. Is your process or design thinking similar to that of another designer?
2. Design is affected by a wide range of influences. Explain how a designer might draw inspiration and solutions in their search for creativity.
3. Many companies now enlist the support of current designers in designing products for the masses. Why are people increasingly searching for products with strong design influences?
4. Select a designer that you have studied and:
   (a) List and describe the main processes used in their work.
   (b) Describe the designer in terms of being a 'design thinker'.
   (c) Identify the factors that ultimately contributed to their success.

Sample HSC questions

Extended-response question A p. 209

1. (a) Designing products that are minimal in their ecological impact is one of today's initiatives for contemporary designers. Identify some of the practical applications used as common principles in sustainable design. (5 marks)
   (b) Describe any of these practical applications that you have used in your MDP. (10 marks)

13 – Trends in design and production

HSC OUTCOMES

A student:	You learn about:	You learn to:
H2.1 explains the influence of trends in society on design and production.	■ trends in designing and producing, including those which are influenced by social, global, political, economic and environmental issues. ■ historical and cultural influences on designing and producing, including – changing social trends – cultural diversity – the changing nature of work – technological change.	■ discuss the issues arising from trends in design and technological activity. ■ identify and acknowledge historical and cultural influences on design and technological development.

13.1 Trends in designing and producing, including those which are influenced by social, global, political, economic and environmental issues

There are many challenges facing companies in bringing new products to the marketplace. One of these is the enthusiasm of people in responding to a perceived need. A large population can collectively form behaviour that can be a decisive trigger in the development of a new design idea or commercial enterprise. What might be seen initially as novel or a craze can take on at large and become a trend, evolving into some permanent change. With the technological and industrial booms design trend has become a phenomenon. The long-term effects on a society can be dramatic long after the fads have ended. The pressure in meeting this social appetite for design products pressures manufacturers to reduce design times, speed up production time, improve quality and monitor competitors.

Successful companies respond to trends influenced by social, global, political, economic and environmental issues; they may do so by altering their industry characteristics strategies and products.

Product customisation

Customer preferences determine the current styles and designs, and producing what customers need with near mass production efficiency, or mass customisation, has become a major trend in industry. Identifying the needs of each customer is a step in understanding the whole market and ultimately develops a framework of mass customisation. To fulfil increasingly sophisticated customer demand, companies must be able to provide a higher degree of product customisation, flexible production systems, and the ability to cope with extended product ranges. A concept known as 'design by customers' is an approach whereby companies communicate to customers what the business can offer in terms of their needs and in assisting them in making choices.

First customers, were you there?

There was a large amount of hype and publicity surrounding Apple's introduction of the iPhone. Whether it is or is not another breakthrough invention, the marketing of this piece has been an advance when compared to Apple's typical product introduction. Apple employees lined up and applauded the first customers who patiently waited for the iPhone's arrival. This definitely personalised the purchasing experience, built up the hype, and reduced the negative aspects of having to stand in line for hours, if not days.

Flexible production systems

The changing nature of consumer demands and the associated evolution of the marketplace have impacted greatly on how products are produced. There is an expectation that production systems be flexible in reducing the time it takes from a product's initial conception to its mass production. This is mainly due to the nature of a shorter product life cycle and the capabilities of rapid prototyping technologies. It is supported by the emergence of sophisticated software, knowledge-based systems and product data management processes.

One of the characteristics of industry transformation is the increasing responsibility and importance of the suppliers. Clear product supply chains requiring high-end transportation logistics are crucial to the production flow. The choice of supplier is very crucial and many become strategic partners within the company. Suppliers are being drawn into the design and development of new systems and components at a very early stage and are more closely integrated to the company's organisational process. There is a higher risk to the supplier now as they take on a greater responsibility within the process. This allows supply companies to adapt products to market and technology changes. The choice of supplier is considered in terms of:

- cost and quality competitiveness
- research and development (R&D) capacity
- proximity to the development centre; the 'just next door' philosophy
- proximity to supply chains.

Flexibility, speed and efficiency of product development result from the way the production and assembly systems are developed. The nature of the design of the facility and the configurations for each production and assembly system are required to remain fixed. Only small modifications are generally made for production volume or product mix changes.

Extended product ranges

The rivalry that exists between companies affects the way they develop strategies. It provides an important impetus for research, design innovations, and changes to the manufacturing process. Companies are vigilant in identifying any changes to consumer preferences, national biases, and new market segments. They need to be flexible by defining variations from each base product. The alternatives found in existing products are how companies can extend the range of that product. While aesthetic details such as colour and shape are obvious inclusions it is the functional design aspects that add to the product's extendibility. All factors in a product can be designed for interchangeability, allowing the extending of the product range in a fast, low-cost way. These new solutions can be tailored to the varied tastes and preferences of global consumers.

SMART car, smart drivers

Figure 13.1 A SMART substitute when requiring an environmentally-friendly design idea

With the focus of the customer towards environmentally friendly and economical substitutes, it seems a perfect time to develop the SMART car. While most customers see it as a replacement to their larger fuel-burning car, the SMART car dealer presents the vehicle as an addition, not a replacement, to your current car, which makes it a very extravagant 'accessory'.

13.1.1 Social issues

Society is a group of interacting communities. There are three types of communities that influence social design issues and bringing new products to market:

- communities of geography relates to a definite location or environment, such as a neighbourhood, suburb, village, town or city
- communities of culture relates to a range of local cliques, subcultures, ethnic groups, religious, multicultural or pluralistic societies and global communities
- communities of need or identity relates to those who are disabled, elderly, and so on as well as broader majority, minority, socio-economic and political groups.

Social issues are related to the fabric of the community, including conflicting interests of community members, and lie beyond the control of any one individual.

Patriotism

People build up their own sense of belonging in society centred on their view of the world and its relationship to their country. It helps form a set of values that are generally held throughout their lives. This means that every Australian will have their own sense of Australian values. Many Australians are reclaiming patriotism by embracing these and redefining what it means to love Australia. The Buy Australian campaign tapped into the patriotism of Australians as a way of supporting Australian manufacturing, its jobs and its profits. 'Buy Australian' is the sentiment of most Australians. Australian states and even local towns will market the quality of their locally-grown produce. In Tasmania people pride themselves on their local produce and manufacturing. Purchasing locally-grown products has spawned a local economy and in addition provided success beyond the island into the global marketplace.

Materialism

Many people see possessions as one of the highest values in life. They surround themselves with various forms of material products as a way to achieve their idea of happiness and satisfaction. Many designers recognise the clear link between these materialistic wants, the gratification of our egos and the substantial amount of products and services available. They respond accordingly by providing expensive products such as yachts, motor vehicles and high-end designer clothing and accessories.

Multiculturalism

A culture is a set of designs for living that is shared by many people. Some of these designs are passed on from generation to generation. Many different cultural groups can mix and form into a multicultural nation. A multicultural society can benefit Australia via ethnic groups adding the richness of their cultures into the Australian cultural mix. A practical aspect of Australia's multicultural policy is that it provides a framework for strengthening community harmony and promotes the economic, cultural and social benefits of Australia's diversity. There are consumer decisions that will influence the products and services they buy. This is based upon the assertion that groups will continue to preserve their culture and will want products and services that are unique to it and the retaining of their history.

Social class

Sociologists tell us that class is just one of the ways that we can divide society, alongside ethnicity, gender, religion or social status. Social class is a way of categorising society by looking at what occupations people have, what they earn, or some other marker of status. Class systems can be used by manufacturers to align products with individualised market selections. In this, design and production groups are able to develop products that provide enough variations to satisfy all people's needs. This idea of increasing the market share enables the company to produce the same type of commodities but with varied styles and price ranges.

13.1.2 Global issues

The concept of a global market incorporates a diverse array of products from all parts of the world in what is called 'global merchandising'. The global market provides a high degree of competiveness for design and production. Many of the design and production processes are via multinational companies. Multinational firms are investing in record numbers in developing nations, providing benefits to those countries via sourcing commodities on a global scale.

Developments and advances in transportation and communication systems allow for business to be conducted on a global basis. These advances have allowed for the rise of global markets where consumers can order products from a wide variety of online shopfronts and receive them within days. With the degree of competition it is easy to see how smaller or developing nations may feel threatened by the wealth and influences of larger, more developed countries.

13.1.3 Political issues

Laws, governments and pressure groups known as 'watch dogs' provide protection to the consumer by maintaining vigilance and responding to the way companies and institutions run their businesses. Legislated law affects the operations of businesses and companies by defining expected procedures of operation. There are also government bodies or agencies set up to closely monitor companies throughout the lifecycle of their designing and producing. These include the following.

- Agencies such as WorkCover determine the safety policies to be put in place in the workplace.
- The Australian Competition Consumer Commission (ACCC), controls unfair competition practices in Australia while the Foreign Investment Review Board controls any foreign involvement.
- Any marketing organisation in Australia must follow the guidelines from the *Trade Practices Act* dealing with consumer protection. The Act covers issues involving anti-competitive pricing and unfair consumer practices as well as product disclosures to the consumer, product security and warranties.
- Standards Australia is a regulatory body that establishes criteria for all design and production to work towards. Any innovation that requires commercialisation must first satisfy a series of stringent controls to qualify as meeting the standards.
- Tariffs are used to control trading prices. Cheaper imports to Australia are given a tariff to protect local markets and therefore the local economy and jobs.
- Quotas also provide protection to local companies as the government can force a quota which sets a limit on how much of a product can be imported into the country.
- Other laws regarding control of imported goods are enforced through customs and quarantine regulations. These are vigilantly monitored and recognised as an arm of Australia's border protection.

13.1.4 Economic issues

The economic environment is affected on a variety of international and national stages and can impact on designing and producing. Global financial crises, climate change, unemployment, lack of economic growth, risk of deflation, poverty and international political problems are key contenders to economic instability. Designers and design companies have been affected by the global financial crisis because government economic initiatives such as providing loans, grants, commissions, share issues, licences and royalties are reduced in number and size.

The economy of scale will impact financially on some businesses as larger companies or conglomerates use 'total buying power' to provide goods and services to a variety of consumers. Total buying power drives down costs to consumers and becomes very competitive in its operation. Consumerism involves consumers developing brand loyalty, resulting in them selecting to purchase brand A over brand B, resulting in ongoing financial support for the producer.

Often companies attempt to drive down costs in order to reward brand loyalty and attract new customers. They may make the decision to move their manufacturing base to places that have a cheaper labour force. This concept of taking the company 'offshore' means that manufacturing costs will be dramatically reduced due to the significantly lower wages of the workforce. While tariffs may be added, the consumer still has access to a cheaper product. However, this can have a crippling effect on the 'home' economy and result in job losses. To counteract this, Australian marketing initiatives support the purchase of Australian-made products. While relatively more expensive, buying Australian will maintain Australia's manufacturing labour force and keep much of the profits in Australia.

13.1.5 Environmental issues

The impact on the natural environment is likely to be the most significant issue confronting Australia and the global community. Environmentally-conscious pressure groups, known as 'green groups', are stepping up their pressure on state and federal governments to make more progress on environmental issues. Governments are realising how unresolved issues of air and water pollution, waste disposal, greenhouse gases and climate change have an affect on the condition of the planet and are working with business, industry and community groups to:

- reduce harmful emissions to air, land and water
- reduce the number of significantly contaminated sites
- reduce the exposure of the community and the environment to chemicals
- develop stringent control laws such as the *Protection of the Environment Operations Act 1997*
- develop pollution prevention and monitoring strategies
- implement best practice of recycling as well as reusing technologies
- pursue carbon neutral strategies such as 'clean energy'
- encourage ecologically sustainable development
- encourage the community to connect with and enjoy the environment in order to maintain and improve their physical and mental health.

For additional information on environmental design refer to Chapters 5 and 14.

13.2 Historical and cultural influences on designing and producing

13.2.1 Changing social trends

The changing nature of society is based on values, groupings, social trends and demographic shifts; a variety of changes have impacted on Australian social trends. Before the arrival of European settlers, Aboriginal and Torres Strait Islander peoples inhabited most areas of the Australian continent. Their lifestyles and cultural traditions differed from region to region and their highly-developed traditions reflected a deep connection with the land. After the arrival of the white settlers in 1788, both convicts and free immigrants began to establish and settle towns and villages.

The roles of men and women have changed since early industrial days and this can be seen in the attitudinal and behavioural changes associated with the current demographic reality. The concept of men as 'metrosexuals', for example, can drive product customisation, leading to businesses adopting practices such as niche marketing. A niche market is a small section of the total market which a company targets with specific products that aim to satisfy that niche's needs and wants.

Traditional social trends	Contemporary social trends	Reasons for changes
Family: large extended families consisted of eight to ten people	Nuclear and single parent families	Birth control invented and accepted Divorce and not being married carries no social stigma Choice of being independent with no children is now an option in society
Families lived in the same house or close by	Families move away from 'home' to other locations in Australia or overseas	Transportation developments mean that people can travel longer distances in shorter times Communication developments mean that long-distance communication is affordable, easily accessed, and can include visual images as well as voice communications, for example, Skype
Vocation: people often followed their parents' vocation	People select their own vocation	In the past, children were raised to complete chores and thus developed skills in areas that translated to vocational skills; today, children are encouraged to look at opportunities further afield as the world comes to them via multimedia such as television and the Internet
Working in the home or nearby	Working outside the home and commuting further or working from home via the Internet	Technological developments in household appliances means that people have more time to spend outside the home as it takes less time to complete household duties Access to broadband communication has presented greater work opportunities and freedom
Lifestyle: units of production	Units of consumption	Moving away from the 'farm' and traditional skills means that most people now purchase what they require rather than producing it, for example, food and clothing
World knowledge: people were often only aware of what was happening in their immediate vicinity	People are more aware of global issues	Internet access has brought the world into people's homes

13.2.2 Cultural diversity

The existence of broad cultural groups within Australian has been part of the nation's modern history. Australian culture is constantly developing and cultural diversity is one of Australia's greatest attributes. Our community and the life we live are much richer as a result of the successive generations of people who have made their home here. Australians come from over 200 countries, and each of these groups has its own cultural diversity as a result of history, ethnicity, language and religion, and value and belief systems. Valuing cultural diversity is the process of recognising, utilising and benefiting from the variety of cultural backgrounds and the associated diversity of traditions and customs, foods, materials, techniques and designs. All these elements have a strong impact on designing and producing. Designers will respond to the market impact based on the consumption of specific products by different cultural groups. This will create niche markets that must be met by designers fulfilling consumer needs.

13.2.3 The changing nature of work

Trends and changes such as globalisation, technological innovation, deregulation and downsizing are generally presented as forces that impact on work environments and the changing nature of that work. These changes are occurring to the nature of work itself and in the composition of the workforce. Looking at the structure of the workforce as a whole, it can be noted that:

- the shifting patterns of the workforce are involving the participation of more women
- unemployment has re-emerged as a significant factor
- there are increasing numbers of workers employed under part-time and casual arrangements
- employment has shifted across industries, with declining jobs in manufacturing and increases in service industries
- the workforce is less unionised.

The nature of work has changed in a number of ways.

- Workers are now expected to work smarter, not harder; many people have fewer co-workers due to computerisation and mechanisation but are still expected to maintain or even increase output.
- Changing work patterns may also alter traditional patterns of living as the boundaries between work and free time become blurred.
- The boundaries between the workplace and the home have broken down, with home becoming the place of work for many due to the availability of access to the Internet.

These shifts constitute a complex restructuring of the world of work and they are helping to drive the practices of designing and producing.

13.2.4 Technological change

Technology is the driving force in our world that intentionally or indirectly causes or accelerates social, cultural or behavioural change. As technology changes it influences every sector of our society on a global basis, as seen in the social networking phenomena of the Internet. Technology provides the forces that break down local and national barriers and enable societies to transact, commute and learn across artificial boundaries.

The technology that drives the information age requires varied practices within designing and producing. Due to the high-end nature of computer technologies, technical expertise is spread across hundreds of specialist companies. What is known as business and manufacturing networks allows these specialist companies to take part in the development of new products and their associated risks and costs. Traditionally, industry started with materials and processed a product from start to finish but the use of high-end technology has changed the nature of production and resulted in the outsourcing of high-tech elements to inter-firms within the production process.

The proliferation of inter-firm networks is shown in Silicon Valley, USA. Here, technology-based companies can flourish in close proximity. The vitality of the region is enhanced through collaboration, which breeds complementary innovation, cross-fertilisation among the inter-firm companies and an ability for production to adapt swiftly to variations such as shorter production cycles or rapid technological change. The involvement of company networking also means the costs and risks of developing new technologies can be shared. It can also foster shared innovation practices among the specialised firms.

Electronic technologies have facilitated the computerisation of certain aspects of the production process, rendering many trade skills and occupations obsolete. Because of this, multi-skilling is essential in most modern industries. Computerisation has increased the need for workers to possess basic clerical and computer skills and the boundaries between many occupations have now become blurred.

Technology has improved society through global trading, provided efficiencies in terms of human productivity, and improved the quality of life for most people in developed countries. Many ethical questions are raised, however, because of negative impacts such as unwanted products, pollution, the depletion of natural resources through unsustainable development practices and the influence on the values of a society when new technology often raises new ethical questions. Philosophical debates will continue over the use of technology in society and will be centred on the harm to the environment, the alienation of people who do not possess the technology and the ethics of new technological practices. Proponents of ideologies such as 'techno-progressivism' see continued technological progress as beneficial to society and the human condition. Whether this is correct or not, technological progress will always impact on the practices of designing and producing.

Key concepts and definitions

Clique—a small, exclusive group of people.

Contemporary—modern, at the present time.

Egalitarianism—the philosophy that all people are equal.

Globalisation creates an international setting, not just a single country. Many companies operate around the world and are therefore recognised as global.

Market share—a strategic management and marketing strategy to envelop the total availability of the market in terms of designing and producing.

Materialism—a desire for wealth and material possessions in the belief that they constitute the greatest good and highest value in life as opposed to ethical or spiritual matters.

Multiculturalism—when many different racial and ethnic groups mix into one nation.

Multinational—a large corporation or company with operations and subsidiaries in at least several countries.

Patriotism—love of and/or devotion to your country.

Pluralistic—a theory that indicates there is more than one basic principle.

Product life cycle—the sequence of stages a new product progresses through from introduction to growth, maturity and decline.

Quintessential—the pure and concentrated essence of a substance or the most perfect embodiment of something.

Quota—a proportional part or share that is fixed by the government to control the amount of imports entering a country.

Rapid prototyping—the automatic construction of physical objects using additive manufacturing technology. It is also known as 3-D printing.

Subculture—a group having social, economic, ethnic or other traits distinctive enough to distinguish it from others within the same culture or society.

Tariff—a system of duties or costs imposed by a government to control the pricing of imports and exports.

Trend—the consideration of a style or preference by people. It generally develops as a line of general direction of movement over time and can be statistically detected.

Useful websites

- rapidbi.com/created/porterfiveforces.html
- www.allbusiness.com/business-planning-structures/business-structures/405782-1.html
- www.bp.com/genericarticle.do?categoryId=98&contentId=7009879
- www.i2.com/supplychainleader/issue4/html/SCL4_short_product_lifecycles.cfm
- www.marketingcentre.com.au/tl_files/mc_files/Articles/P&L%20Magazine%20article%20-%20The%20Eureka%20principle.pdf
- www.nickwebb.com/Keys1.html
- www.teleocracy.com/Chapter%201.pdf
- www.teradata.com/tdmo/v08n04/Features/CoverStory/InnovationStrategy.aspx

Classroom activities

1. Examine how global technology development affects you as a consumer. For example, discuss the concept of global consumerism.
2. Develop a campaign model that you could use to show the government how to solve the problem of electronic waste. For example:
   - recognise the types of electronic products and their life cycles
   - target advertising towards specific user groups and/or particular items
   - saturate media with associated global problems and their solution
   - use legislation to ban environmentally dangerous products
   - develop a mandatory product return legislation placed on retailers to ensure tracking of each product sold for their end cycle return
   - place a refundable deposit on items that the consumer returns to the retailer
   - provide financial incentives to companies based on usage of recycled materials in their products
   - develop a plan to de-engineer the expended products
   - link in with celebrations such as World Environment Day and Earth Day.
3. Trace back your family history and define some of the technological moments in your ancestors' lives.
4. Identify an ethical question that new technology raises and discuss its implications for today and for future concerns. For example, impacts could relate to current issues that are not likely to be addressed in the near future.

Key concept questions A pp. 209–210

1. Why is it necessary to understand customers' preferences when designing and producing?
2. How are computer and communication companies like Apple able to produce so much 'hype' with each new product release?
3. List the elements that impact on a flexible production system.

4. Within the Venn diagram add detail to show how the communities are connected.

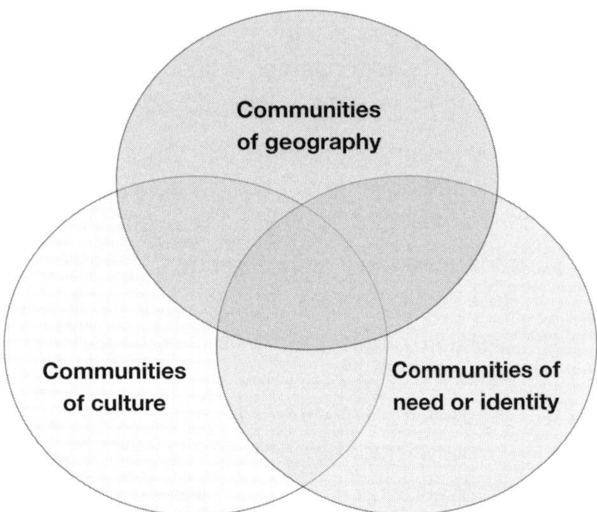

5. Describe the role of the Australian Competition Consumer Commission, the Foreign Investment Review Board and Standards Australia in terms of monitoring companies.
6. Why is it now an imperative for Australian companies to move offshore?
7. Describe what motivates trends in designing and producing.

Sample HSC questions

Objective-response questions A p. 210

Circle A, B, C or D. (1 mark each)

1. The evolution of computer technology has led to many changes in traditional work practices. What has impacted the most on the workplace?
   A Much of the work is done from home.
   B Workers spend too much work time involved in web-based entertainment.
   C Most workers need to possess basic clerical and computer skills.
   D Workers do not talk as much at work now.
2. With the advent of digital television many homes needed to dispose of their old analogue sets. When planning a technology change, designers must ensure that
   A they do not oversupply the market.
   B the components are able to be disassembled for recycling.
   C they trade with other countries and sell the old technology.
   D a rubbish tip is available.
3. Which is the most influential trend on design and production?
   A the size of the market
   B the employment patterns of the workers
   C the image the company presents to the wider community
   D the skill of the workforce

Ethical and environmental issues

HSC OUTCOMES

A student:	You learn about:	You learn to:
H2.2 evaluates the impact of design and innovation on society and the environment.	■ ethical and environmental issues – ethical and environmental considerations for designers and society – sustainable technologies – protection of intellectual property – rights and responsibilities of the designer – impact on Australian society.	■ critically analyse ethical issues in relation to innovation. ■ discuss ethical and environmental considerations for designers and society in general. ■ identify the factors which contribute to the efficiency and sustainability of technologies.

This chapter builds on information provided in the Preliminary course. It is recommended that you revise Chapter 5 before progressing further.

14.1 Ethical and environmental issues

The design industry is constantly evolving and changing under the influences of culture, trends, innovative technology and the economy. Some designers ignore their moral and ethical roles in society, and the responsibilities that come with these, by focusing more on business, marketing and corporate clients than on ethics and social responsibility.

14.1.1 Ethical and environmental considerations for designers and society

Design is an ever-increasing part of every person's life. It is about creative problem-solving using innovation. Designers are to the information age what engineers were to the age of steam and what scientists were to the age of reason.

Because most designers have corporate clients they can influence the impact these corporations have on the world. The majority of design projects address corporate needs with an over-emphasis on the commercial sector of society, which consumes most of designers' time, skills and creativity. A designer's clients play a large role in how they are perceived and what kind of work they do.

In many countries consumers, influenced by the media and their physical, social and emotional needs and wants, spend more and more money on products. With more consumption comes an increase in waste and pollution. Designers need to be aware of the practices of their clients as well as their role in continuing the cycle of consumerism.

Many leading designers have begun to question the role they play in society. Other designers argue that design is just a job and ethics does not need be a priority. In one way or another, design ethics has weighed on the conscience of the design industry. Some are truly making a difference, making social and ethical design their life's work. One method for achieving this is the concept of 'true cost' design. This means that before designers create a new product they consider the ecological and psychological consequences of their design.

Design requires a code of ethics where the impact and importance of design is acknowledged, and the harm it can do reduced. In 2001 the Graphic Designers of Canada Association adopted a code of ethics and professional conduct. The code established guidelines and responsibilities for graphic designers in relation to their clients and issues such as the environment, society and human rights. The code was a bold movement towards ethics as it suggests that designers should refuse to work for clients that engage in unethical practices, including doing environmental harm or disregarding human rights. Designers who fail to uphold the code risk losing their status as registered graphic designers.

Design plays a big role in consumerism, globalisation, branding and in sustaining the environment, thus ensuring light global carbon footprints. To achieve this, designers must examine their role in the world. As talented, creative people they have the potential to be positive influences in society and must take responsibility for the work they produce.

14.1.2 Sustainable technologies

Sustainable design, otherwise called environmental design, environmentally sustainable design (ESD) or environmentally-conscious design is the philosophy of designing physical objects, the built environment and services to comply with the principles of economic, social and ecological sustainability. The aim of sustainable design is to eliminate negative environmental impacts through skilful and sensitive design. Products of sustainable design require no non-renewable resources and have a minimal impact on the environment.

Sustainable design is a reaction to the global environmental crises, the rapid growth of economic activity and human population, the depletion of natural resources, and damage to ecosystems and loss of biodiversity.

Sustainable technologies have a light ecological footprint on the planet. They use less energy and fewer limited resources. In addition, they do not deplete natural resources or pollute the environment directly or indirectly. They can be reused, renewed or recycled at the end of their life and often promote a cradle to cradle approach in their life cycle.

Appropriate technology emphasises the suitability of the chosen technology. Often the most appropriate technology may not be the most sustainable one.

The limits of sustainable design in reducing ecological impacts are considerable because consumption is consistently outpacing production. The present approach, which focuses on the efficiency of delivering individual goods and services, does not solve this problem.

Further information on how to achieve sustainable design and technology can be found in Chapter 11.

14.1.3 Protection of intellectual property

Intellectual property (IP) represents the property of your mind or intellect. It can be an invention, trademark, original design or the practical application of a good idea and is often the edge that sets successful companies apart. As world markets become increasingly competitive, protecting intellectual property becomes essential. Confidential information (also referred to as trade secrets), patents, registered designs, trademarks, copyright, circuit layout rights and plant breeder's rights are all legally classified as IP rights.

Inventions can be protected by a granted patent. It is a effective protection right granted for any device, substance, method or process which is new, inventive and useful. It is legally enforceable and gives the owner the exclusive right to commercially exploit the invention for the life of the patent. In return, patent applicants must share their knowledge by providing a full description of how the invention works. This information becomes public and can provide the basis for further research by others.

There are two types of patents in Australia: a standard patent gives long-term protection and control over an invention for up to twenty years and an innovation patent is a relatively fast, inexpensive protection option, lasting a maximum of eight years.

To obtain a patent, the invention must be a new type of manufacturing technique, novel, useful and possess a sufficient level of inventiveness. The types of inventions may include products and devices, mechanical apparatus, business methods, biotechnology-related products, and computer- and Internet-based software and processes.

You cannot patent artistic creations, mathematical models, plans, schemes or other purely mental processes. Australian patents are administered by the Patent Office of IP Australia, a federal government organisation.

Obtaining a granted patent adds value to your invention and your business in the following ways.

- You can benefit from the financial rewards.
- You can bring patent infringement proceedings against an unauthorised user of the invention, protecting its monopoly and market share.
- Having a patented invention will assist in forming relationships with potential partners, venture capitalists and licensees. It demonstrates to the potential business partner the viability, value and potential market share claimed by the patented invention.
- A patent protects the investment in time and resources that you have put into the research and development.

The patent system is structured to enable you to make this decision before the major costs of patenting are encountered. For example, filing a provisional application is quite inexpensive and it gives you twelve months to consider the commercial worth of your invention and to resolve issues such as finance and licensing. There are a number of different types of patents such as standard, innovation and international patents.

A trademark is a sign used by a business to set its activities apart from those of its competitors. Some famous examples include Nike's 'swoosh' and McDonald's golden arches. If you are thinking about a new product or service and you want to establish an image for it, you should also be thinking about a distinctive trademark.

Trademark clearance searches are used to ensure that your trademark is available for registration, and check that you will not infringe on other traders with existing marketplace rights. A trademark is not registrable if it is not capable of distinguishing your goods or services from those that are the same or similar. Trademarks which conflict with an earlier trademark, or would mislead the public about the nature of the goods and services, are also difficult to register.

In addition, some words are protected by law and cannot be registered as trademarks. Some are prohibited as trademarks under the *Trademarks Act 1995*, for example, 'Olympic Champion'.

Copyright protects literary, dramatic, musical and artistic works. Unlike other forms of intellectual property in Australia, copyright is not conferred by registration but rather automatically arises upon the creation of the original work in material form. This automatic right generally applies for the life of the creator and then a further seventy year period. Commercial copyrighted materials can include manuals, marketing materials, text books, Internet and website content, architectural plans, artworks, music, lyrics and software codes.

A copyright infringement, breach or violation can arise if a competitor or employee, without your authority, reproduces or takes a substantial portion of your work. Authors and creators of copyrighted work also have moral rights. These rights can be infringed where authorship is not recognised on a work or the work has been modified or displayed in such a way as to be considered derogatory.

With the commercial growth of the Internet, the value in domain names has increased significantly. With this has come issues such as domain name disputes and 'cyber squatting' where an unauthorised party registers the domain name which is identical or closely similar to another businesses trademark or trading name. The domain squatter's predominant purpose is to derive commercial gain.

14.1.4 Rights and responsibilities of the designer

Environmental rights and responsibilities work together. We have the right to a clean environment but if we do not act on our responsibilities in terms of recycling waste and so on then this right will not become a reality.

Water

Adequate water is necessary for agriculture, manufacturing and human consumption.

Rights include the following.

- Individuals have the right to access clean water now and in the future.
- Communities have the right to environmental education and clean accessible water supplies.
- Businesses have the right to use the water they need and to invest in environmental issues.
- Governments have the right to make laws to protect water for human and non-human use and to set consequences for breaking those laws.

Responsibilities include the following.

- Individuals have the responsibility to use water efficiently and to avoid polluting it.
- Communities have the responsibility to educate themselves and to report problems with water usage.
- Businesses have the responsibility to invest in better practices regarding water consumption.
- Governments have the responsibility to hold individuals, communities and businesses accountable for their actions.

Sustainable resources

Sustainability must remain economically viable in order to protect and restore the natural environment.

Rights include the following.

- Individuals have the right to use environmental resources and to protect the ecosystem and their health.
- Communities have the right to access clean air, water and land.
- Businesses have the right to be economically viable.
- Governments have the right to choose the best way of using natural resources in a manner that is healthy for the community and the environment.

Responsibilities include the following.

- Individuals have the responsibility to plan for the future and to promote a healthy environmental atmosphere.
- Communities have the responsibility to sustain and improve their environment.
- Businesses have the responsibility to pursue environmentally-friendly conduct.
- Governments have the responsibility to respect the opinions of the community, provide resources to community members and to implement laws, regulations and incentives that promote best management practices.

Renewable and non-renewable resource management

Waste reduction and renewable resources must be linked to promote environmental equity.

Rights include the following.

- Individuals have the right to environmental education so that they are aware of the importance of recycling, reusing, and buying repaired/used products.
- Communities have the right to be informed about the environment, to be active on behalf of their environment and to access recycling facilities.
- Businesses have the right to decide about the type of resources they use, how they use them, and their environmental plans.
- Governments have the right to manage their own resources based on national needs.

Responsibilities include the following.

- Individuals have the responsibility to use their environmental education to their advantage and to not waste valuable resources.
- Communities have the responsibility to be informed about the environment and participate in environmental activities, including recycling.
- Businesses have the responsibility to minimise pollution and waste of natural resources and to replace and neutralise everything they waste in order to maintain a resources balance.
- Governments have the responsibility to use and encourage the use of sustainable resources and to inform constituents about usage of sustainable resources.

Ethics and stewardship

Ethics and stewardship help ensure community economics and environmental health. They can have long-lasting and positive impacts on communities.

Rights include the following.

- Individuals have the right to preserve nature and enjoy biodiversity.
- Communities have the right to be educated in order to preserve nature.
- Businesses have the right to use nature to make a profit.
- Governments have the right to make regulations concerning clear water, air and land.

Responsibilities include the following.

- Individuals have the responsibility to not harm the environment and to replace or fix what they damage or destroy.
- Communities have the responsibility to not overuse resources.
- Businesses have the responsibility to not use more resources than is necessary.
- Governments have the responsibility to protect our environment.

14.1.5 Impact on Australian society

The following Australian Government website provides up-to-date information on environmental impacts on Australian society.
australia.gov.au/topics/environment-and-natural-resources

Some of the departments and organisations dealing with environmental issues are listed below.

Air and weather

- Atmosphere relates to initiatives to improve the quality of air and to ensure the recovery of the ozone layer. Includes links for information on air quality, fuel alternatives and greenhouse gases.
  www.environment.gov.au/atmosphere/
- The Australian Government Bureau of Meteorology (BOM) provides weather forecasts, warnings and observations for all states and territories of Australia. There is also information about climate, hydrology and other weather services.
  www.bom.gov.au/
- The Centre for Australian Weather and Climate Research is an equal partnership between the Bureau of Meteorology and the CSIRO.
  www.cawcr.gov.au/
- The Bureau of Meteorology climate information pages contain additional maps, educational materials and access to services for agriculture.
  www.bom.gov.au/climate/
- The Managing Climate Variability Program invests in research and development activities to improve the information and tools available to Australian farmers and natural resource managers for the management of climate-related risks.
  www.managingclimate.gov.au/

Climate change

- The Department of Climate Change and Energy Efficiency delivers the Australian Government's climate change strategy.
  www.climatechange.gov.au/

- The Department of Climate Change also publishes Australia's National Greenhouse Accounts, which assists the government in developing climate change policy and setting emissions targets. It is also used to track progress against Australia's targets under the Kyoto Protocol.
  www.climatechange.gov.au/climate-change/emissions.aspx
- The Carbon Pollution Reduction Scheme outlines the final design of the scheme and the medium term target range for reducing carbon pollution.
  www.climatechange.gov.au/publications/cprs/white-paper/cprs-whitepaper.aspx
- Cities for Climate Protection (CCP) Australia provides information to local governments and their communities on ways to reduce greenhouse gas emissions.
  www.environment.gov.au/settlements/local/ccp/
- Climate Change in Australia has developed climate change projections in order to understand the magnitude of climate change and its potential impacts.
  www.climatechangeinaustralia.gov.au/
- The Climate Ready Program provides grants on a matching funding basis to support research and development for solutions to climate change challenges.
  www.ausindustry.gov.au/InnovationandRandD/ClimateReadyProgram/Pages/ClimateReadyProgram.aspx
- The National Carbon Accounting System (NCAS) provides a complete accounting and forecasting system for human-induced sources and sinks of greenhouse gas emissions from Australian land-based activities.
  www.climatechange.gov.au/government/initiatives/national-carbon-accounting.aspx

Ecosystems

- The CSIRO Centre for Arid Zone Research is engaged in research projects studying arid and semi-arid ecology and natural resource management.
  www.cazr.csiro.au/
- The CSIRO Sustainable Ecosystems is focused on maintaining the sustainability of Australia's landscapes, environments and communities. Their research targets a diverse range of regions and sustainability issues.
  www.cse.csiro.au/
- The CSIRO Tropical Ecosystems Research Centre researches issues such as conservation management of natural and semi-natural areas, tropical agriculture and horticulture and ecology, and integrated control of weeds.
  www.terc.csiro.au/
- The CSIRO Tropical Forest Research Centre provides a research base for a number of scientists engaged in projects studying aspects of the tropical forest environment.
  www.tfrc.csiro.au/
- Antarctica is the world's greatest wilderness, a legacy of the past supercontinent, Gondwana. Although one of the harshest environments on the planet Antarctica is also one of the most vulnerable, and one which Australia and other parties to the Antarctic Treaty are committed to protecting.
  www.aad.gov.au/default.asp?casid=42
- The Wet Tropics Management Authority is a small agency responsible for managing the Wet Tropics World Heritage Area.
  www.wettropics.gov.au/

Energy

- CSIRO Energy Technology looks at the ways in which the generation and consumption of energy and power affects Australia and Australians. They investigate ways in which processes can be made more efficient and better for the environment.
  www.csiro.au/org/ET
- The Energy Reform Implementation Group (ERIG) investigated how best to encourage the development of a national electricity transmission system. This page links to the executive summary and main report 'The Way Forward for Australia'.
  www.ret.gov.au/energy/energy_markets/Pages/erig.aspx
- The development of a white paper on energy issues was released in late 2009.
  www.ret.gov.au/energy/facts/white_paper/Pages/energy_white_paper.aspx
- The National Framework for Energy Efficiency (NFEE) aims to unlock the economic potential associated with the increased uptake of energy efficient processes and technologies across Australia.
  www.ret.gov.au/Documents/mce/energy-eff/nfee/default.html

- The Office of the Renewable Energy Regulator (ORER) is a statutory authority established to oversee the implementation of the government's mandatory renewable energy target.
  www.orer.gov.au/
- Through its Renewable Energy Target (RET) Scheme, the government has pledged that by 2020, twenty per cent of Australia's electricity supply will come from renewable sources. In ten years the amount of electricity coming from sources such as solar, wind and geothermal will be around the same as all of Australia's current household electricity use. The RET expands on the existing Mandatory Renewable Energy Target (MRET), which began in 2001.
  www.climatechange.gov.au/government/initiatives/renewable-target.aspx

Environmental impact

- The Department of Climate Change publishes data that assists the government in developing climate change policy and setting emissions targets.
  www.climatechange.gov.au/climate-change/emissions.aspx
- The Department of Climate Change and Energy Efficiency delivers programs under the government's climate change strategy.
  www.climatechange.gov.au/
- The National Carbon Accounting System (NCAS) provides a complete accounting and forecasting system for human-induced sources and sinks of greenhouse gas emissions from Australian land-based activities. Around twenty-seven per cent of Australia's human-induced greenhouse gas emissions come from activities such as livestock and crop production, land clearing and forestry.
  www.climatechange.gov.au/government/initiatives/national-carbon-accounting.aspx
- Search the National Pollutant Inventory (NPI) for free information about substance emissions in Australia.
  australia.gov.au/service/pollutant-inventory

Environmental sustainability

- The ESD Design Guide for Office and Public Buildings gives a basic introduction to ecological sustainability issues and specifically how the built environment affects them.
  www.environment.gov.au/settlements/government/publications/esd-design/index.html
- The government's LivingGreener website provides a starting point for information about living more sustainably and reducing your environmental impact.
  www.livinggreener.gov.au/
- The National Chemical Reference Guide (standards in the Australian environment) is designed to help you find relevant information about chemicals as quickly and easily as possible.
  www.environment.gov.au/chemicals-guide/
- The National Program for Sustainable Irrigation focuses irrigation research on critical emerging environmental issues while also aiming to improve the productivity of irrigated agriculture and maximise community benefits.
  www.npsi.gov.au/
- Information about sustainable transport including fuel consumption labels, fuel consumption guides, alternative fuels and travel demand management can be found at
  www.environment.gov.au/settlements/transport/index.html
- TravelSmart Australia is about reducing our reliance on cars and making smart choices about other forms of transport.
  www.travelsmart.gov.au/
- The Water Efficiency Labelling and Standards Scheme (WELS) aims to improve water efficiency by assisting people to purchase and industry to manufacture more water-efficient products.
  www.waterrating.gov.au/

Pollution and waste management

- Visit the hazardous waste home page from the Department of the Environment, Water, Heritage and the Arts.
  www.environment.gov.au/settlements/chemicals/hazardous-waste/
- Find out about emissions in your area and search the National Pollutant Inventory (NPI) for free information about substance emissions in Australia.
  australia.gov.au/service/pollutant-inventory
- Taking action to reduce waste by encouraging material efficiency, reducing the generation of waste and enabling the recovery and reuse of discarded material is a critical element of sustainable development.
  www.environment.gov.au/settlements/waste

Key concepts and definitions

Design ethics links consumerism to the environment, perhaps by adopting 'true cost' design.

Environmental issues or environmentally sustainable design (ESD) is the philosophy of designing products, systems and environments that comply with the principles of economic, social and ecological sustainability.

Intellectual property (IP)—the property of your mind or intellect. It can be an invention, trademark, original design or the practical application of a good idea. Under law, intellectual property can be protected.

Patents are a means of protecting an invention. A patent is a legally enforceable right granted for any device, substance, method or process which is new and inventive.

Useful websites

- environment.about.com/od/greenlivingdesign/Environmental_Issues_Green_Living_Design.htm
- graphicdesign.about.com/od/copyright/f/own_rights.htm
- www.c-risk.com/Articles/sgms_s&npath.htm
- www.designmatrix.com/services/environ.html
- www.ipaustralia.gov.au/
- www.wipo.int/about-ip/en/

Classroom activities

1. Explain how you could legally protect the intellectual property of your MDP.
2. Describe your responsibilities in terms of your MDP.
3. What are your rights if someone else stole your idea and patented it under their name?
4. Discuss who owns your MDP intellectual property. Is it you, your mentor, your teacher, the school or someone else?

Key concept questions A p. 211

1. Define the term 'intellectual property'.
2. Describe what is meant by the term 'patent'.
3. Explain what is meant be the term 'design ethics'.
4. Differentiate between the concepts of 'reuse', 'renew' and 'recycle'.
5. Suggest and justify what 'design impact' should measure.

Sample HSC questions

Extended-response questions A p. 211

1. Critically analyse the importance of ethical and environmental design.
2. Explain what is meant by 'sustainable design' and explain how each of its principles is used to promote sustainability.
3. Describe what is meant by the term 'intellectual property'. Explain how patents can be used to protect the different types of intellectual property.

(15 marks each)

actors that influence innovation and its success

HSC OUTCOMES

A student:	You learn about:	You learn to:
H3.1 analyses the factors that influence innovation and the success of innovation.	■ factors that impact on success of innovation including – timing – available and emerging technologies – historical, cultural and political factors – economic and legal factors – marketing strategies – the role of a variety of agencies that may impact upon the success of innovation.	■ differentiate between factors which have contributed to the success or failure of innovations. ■ describe the role of a variety of agencies that influence the development, implementation and acceptance of innovation.
	■ entrepreneurial activity – nature of entrepreneurial activity – role in design and technological activity – agencies which influence entrepreneurial activity – management and entrepreneurial activity – legal and ethical issues.	■ discuss the influence of entrepreneurial activity on successful design and innovation. ■ discuss the legal and ethical issues related to entrepreneurial activities.

15.1 Factors that impact on success of innovation

A key factor in today's economic world is creating a sustained advantage ahead of the competition. To do this, many companies need to deliver new products and services through a creative process known as innovation. While an invention is the process of discovering or creating something new, an innovation is often based on an invention that is then changed or improved to suit a changing need. The concept behind entrepreneurialism is to move the innovation into the marketplace in order to make a profit.

Australia's future depends upon effective strategies that drive successful innovations. The most productive and fastest-growing economies are developing innovative capabilities. If they do not, it is unlikely that they will be able to maintain their existing levels of long-term economic growth in the face of new dynamic competitors.

Factors that are used to gauge the impact on success include timing, available and emerging technologies, cultural, political, economic and legal factors, and marketing strategies including size, demand and product promotion. While seen as individual these factors are generally causal in the way they often interrelate. More information on this can be found in Chapters 2 and 11.

15.1.1 Timing

The timing of an innovation usually occurs when management or a consumer perceives a need. All companies recognise that at any time a change of circumstances might occur from changes in government policy, pressure from public opinion groups or economic forecasts. It is important that companies develop mechanisms to plan for the future.

Through this, companies can confidently predict the 'right time' by analysing market forces likely to shape the ideal time for the innovation's launch. This management technique is known as strategic planning and is also used in building capacity in risk management and contingency planning. It is an organisation's way of defining its strategy and direction and pursuing expected goals.

15.1.2 Available and emerging technologies

Technological discoveries or inventions can provide a stimulus to initiate the innovation process. Consistently successful companies in innovation place great importance on managing and investing in information systems to obtain reliable and current data about markets, technologies and research.

Technological discovery is the engine of innovation and competitive advantage with understanding and promoting the use of technology being the realm of research and development (R&D) managers. Research and development generally looks at applying high-end technology to manufacturing industries.

15.1.3 Historical, cultural and political factors

External factors such as historical features, cultural mores or political factors can be very important forces driving the success of an innovation. This is seen in the way that a perceived want can fulfil the expectations and associated needs of particular consumer groups. If a design reflects the attitudes of a user group in a particular way then it is highly likely that it will be accepted by that group. Many politicians tap into public opinion and use it to guide policy. In addition, pressure groups can be very persuasive. Innovation is very reliant on government policy as a driving force for future development.

All decisions are based on an individual's perceptions from their personal and historical experiences and their values. Marketing tools are important mechanisms in defining social and cultural roles and expectations in relation to innovations.

15.1.4 Economic and legal factors

External factors such as the economic and legal factors involve laws, government agencies and pressure groups. The government offers a framework of control with the production of goods and services, the distribution of income, the regulation of markets and firms and the management of the economy. Inflation, depression, unemployment, poverty and international political problems can all be precursors to social change. While governments are required to support the development of innovation, the economy must be buoyant enough to justify investing in innovation. Funding an innovation in tight economic times may be considered risky if it requires a considerable investment of time and money. Government initiatives may support the innovation in the short term by providing loans, grants, commissions, share issues, licences or royalties. In times of unstable economies, individual investment and spending habits will change.

Laws attempt to maintain the right balance in relation to developing innovations. The most common associated laws focus on contracts and agreements, and issues such as trade secrets, intellectual property, patents, design registrations, trademarks and copyright.

15.1.5 Marketing strategies

The marketing strategy is the single biggest factor which determines the success or failure of an innovation on its path to commercialisation. The simple mantra in advertising is simple: 'get it right and you prosper'.

Figure 15.1 Consider the factors when marketing an innovation.

Marketing may be defined as 'the process of planning and executing the conception, pricing, promotion, and distribution of ideas, goods, and services to create exchanges that satisfy individual and organisational objectives'. A variety of strategies is used by marketers to help guide how, when and where product information is presented to consumers when the goal is to persuade them to buy a particular innovation. For additional information on marketing refer to Chapter 7.

15.1.6 The role of a variety of agencies that may impact upon the success of innovation

For Australia to succeed in the twenty-first century a culture of innovation needs to be developed. Significant elements in this may include:

- connections and collaborations
- skilling and tooling
- regulation
- business structure and governance
- investment
- market knowledge
- global competitiveness
- infrastructure.
- responsiveness and flexibility.

Government agencies and allied bodies have established support for the development and adoption of new innovations through a range of actions. Many of these agencies aim at enhancing the innovation's productivity and commercial success and also provide structured support. These agencies can also play a key role in helping small enterprises participate in local and overseas markets.

One of the driving forces in innovation is the interaction among researchers; as a consequence companies tend to cluster in the same geographic area. Geographic proximity is conducive to design development: in Europe, watchmakers have clustered in Switzerland, in the US car manufacturing clusters around Detroit, Hollywood for motion pictures and Silicon Valley for electronics, while the four centres of fashion are New York, Paris, Milan and London.

15.2 Entrepreneurial activity

15.2.1 Nature of entrepreneurial activity

An entrepreneur can organise and manage an enterprise or venture and take it through to commercial success, often by identifying a market opportunity and incubating ideas and organising resources to turn it into a successful product. There is assumed risk throughout all development stages and significant accountability for the entrepreneur. Because of this entrepreneurs are seen as being willing to accept a high level of personal, professional or financial risk to pursue an opportunity.

The nature of entrepreneurial activity requires an entrepreneur to have the attitude and drive to succeed in business. All successful entrepreneurs have similar ways of thinking and posses several key personal qualities such as those outlined below.

- Entrepreneurs are driven to succeed and expand their business. They see the 'bigger picture' and are often very ambitious and self-motivated.
- Entrepreneurs set massive goals for themselves and stay committed to achieving them.
- Successful entrepreneurs have a high though healthy opinion of themselves and often a strong and assertive personality.
- All entrepreneurs have a passionate desire to do things better. Entrepreneurs know the importance of keeping on top of their industry and that the only way to stay number one is to evolve with the times.
- Successful entrepreneurs thrive on competition.
- Entrepreneurs are always on the move, full of energy and highly motivated.
- Innovative entrepreneurs are often at the forefront of their industry. They readjust their path if criticism is constructive and useful to their overall plan. True entrepreneurs are pioneers, comfortable fighting on the frontline.

Michael Dell: innovator of the personal computer industry

Michael Dell revolutionised the personal computer industry by innovating a process to mass-produce individually made-to-order computers. He was able to commercialise this innovation by providing customers with a direct online service. Selling directly meant the industry skipped the 'middle man', with the resulting savings being passed down to consumers in reduced prices.

His entrepreneurship in manufacturing and distribution has made him one of the most successful businessmen of our time. Starting out in the dorm room of his university and with just $1000, Michael Dell is now chairman and CEO of a global company with a net worth of over $30 billion, employing over 40 000 people in more than 170 countries. Dell's product line has diversified to including MP3 players and televisions, plus a wide selection of computer services. Dell is the largest online computer retailer, selling an average of $30 million of products a day.

Richard Branson: the 'rebel billionaire' and Virgin

Richard Branson's innovative idea was simply to deliver old products and services in new ways. He developed a strong consumer focus from the start by redeveloping businesses so that customers were served better. Beginning with a record shop named Virgin on Oxford Street in London he expanded this to a successful record label of the same name in 1972. Expansion of his company grew to include Virgin Atlantic Airways, where Branson pioneered services such as seat-back personal televisions. Virgin has continued to diversify its interests with high-tech trains and service and credit cards. In a move to appeal to the youth demographic, the no-contract Virgin Mobile was launched. Branson's most innovative move to date is Virgin Galactic, which flies into suborbital space, offering customers the chance to experience weightlessness for seven minutes on an up-scaled version of Spaceship One.

Anita Roddick: redefining business with The Body Shop

Anita Roddick's innovative idea was a cosmetic store that used only natural ingredients. She was well known for her involvement in activism and campaigned for environmental and social issues. She founded The Body Shop in the early 1970s and it went on to become one of the world's most successful retailers of cosmetics and related products. Following Roddick's own values, The Body Shop has a reputation for environmental responsibility. The company is against animal testing and also supports community trade, human rights and a number of other causes. Today The Body Shop has around 2400 stores and more than seventy-seven million customers in sixty countries. The business sees itself as a lot more than just a beauty company.

15.2.2 Role in design and technological activity

As you work on your MDP you will progress through a series of phases quite similar to that of an entrepreneur in the way they:

- have great ideas and do something about them
- hunt for opportunities to promote their ideas
- thrive on the challenge of creating their own big breaks

- make money out of solving problems and selling their ideas
- usually start up their own businesses.

Continue to research by exposing yourself to as many influences that you can find by:

- talking to your friends
- going to trade shows
- searching for opportunities
- keeping an eye on the latest trends
- writing down your goals
- researching your industry or technology
- talking to people in your field
- keeping focused and not giving up!

Entrepreneur goes mainstream

Figure 15.2 Mark Olive: using the right ingredients for success.

Aboriginal chef Mark Olive is bringing the passion to Indigenous cuisine. For the past thirty years television chef and Bundjalung man Mark Olive has been on a mission to have native herbs, spices and meats adopted by mainstream Australia. 'As we have embraced every other culture's food in this country, it is time to develop our own beyond that of meat pies and Fosters', he says.

Mark Olive labels his food as 'contemporary Indigenous' with dishes such as wallaby lasagne, kangaroo burgundy pie and wattle seed cheesecake being visually appealing. The success of his cooking show *The Outback Café*, and his book of the series, as well as his occasional role as a globetrotting ambassador for Indigenous cuisine, have all helped to push the cause a little further.

Mark Olive is now developing a gourmet experience with his cuisine and is marketing the idea into mainstream menus around 'blue-ribbon' locations in Victoria. Mark Olive believes the time is right for Indigenous food as people more than ever have a willingness to try out new things. he growth in the native food industry also provides much-needed opportunities for younger Indigenous peoples, including an exchange of knowledge and ideas regarding health and nutrition.

Like any entrepreneur, Mark Olive can recognise the development of a chain of systems working from the supplier right back to the Indigenous producers providing ingredients such as wattle seed, lemon myrtle and river-mint. There are currently several native foods market niches being exploited, including minimally processed and further valued-added products, with 'bush tucker', 'bush-foods', 'native foods' or 'Australian cuisine' featuring in tourism guide books.

15.2.3 Agencies which influence entrepreneurial activity

Some of the agencies which influence entrepreneurial activity are outlined below.

AusIndustry: AusIndustry is an independent body established by the Australian government to assist with the administration of its innovation and venture capital programs. Business programs include innovation grants, tax and duty concessions, small business development, industry support and venture capital.
(www.ausindustry.gov.au)

Innovation Australia: this board supports programs involving innovations that are developing renewable energy and green technologies.

Standards Australia: a regulatory body that establishes criteria that all products and processes are required to work towards. Highly recognised by the government, any innovation that requires commercialisation must first satisfy a series of stringent controls to qualify as meeting the standards. A motorcycle helmet is an example of the power of standards in Australia: a motorcycle helmet is not legal unless it displays its standards sticker on the outside. The role of Standards Australia is to set limits and standardisations where processes and products come under their control.
(www.standards.org.au)

australiandesign.org: an initiative of the Australian International Design Awards aimed at

supporting, promoting and enriching the Australian industrial design community. It provides a valuable resource to designers, business, industry, governments, educational institutions and consumers.
australiandesign.org.au

IP Australia: the Australian Government agency responsible for intellectual property including administering patents, trademarks and plant breeder's rights. By granting these rights, IP Australia contributes to the improvement of Australian and international IP systems. As global markets become increasingly competitive it is important for designers to secure their intellectual property.
www.ipaustralia.gov.au

Business.gov: deals with matters concerning businesses at the start-up stage and provides advice and support on what type of business structure, business and marketing plans and export strategies should be used.
www.business.gov.au

The Council of Small Business of Australia: the peak body of small business organisations, industry groups and individual firms in Australia. The council provides a variety of individual support packages such as political and government representation, small business sector information and membership to associations.
www.cosboa.org

Australian Securities and Investments Commission: ASIC is one of Australia's corporate, markets and financial services regulator. To attract confident and informed investors it is important that Australia has a strong economic reputation. ASIC delivers this by ensuring that Australia's financial markets are fair and transparent.
www.asic.gov.au

The Australian Competition and Consumer Commission: the ACCC is an independent statutory authority. It was formed in 1995 to administer the *Trade Practices Act 1974* and other Acts. Its primary responsibility is to ensure that individuals and businesses comply with the Commonwealth competition, fair trading and consumer protection laws. Currently it has become the 'watchdog' for business practices by taking legal action over anti-competitive conduct though it also provides alternatives to litigation. The ACCC is the only national agency dealing generally with competition matters and codes of conduct and the agency with the responsibility for enforcing the *Trade Practices Act* and the state/territory application legislation.
www.accc.gov.au

NRMA Motoring & Services: one of the many non-government agencies prepared to commit to innovation and new products. Companies such as NRMA see innovation as part of their investment strategy and a way forward. Companies that have an insurance arm are generally very receptive to future investment. Innovations may be linked to risk management and therefore reduce insurance payouts, an obvious advantage to insurance-styled companies. It also advantages consumers via reduced premiums. NRMA Motoring & Services invests a great deal in road-related issues, believing that any product that reduces accidents, saves lives and cuts premiums is a benefit to society. The Optalert technology examined in Chapter 18 is a good example of this.
www.nrma.com.au

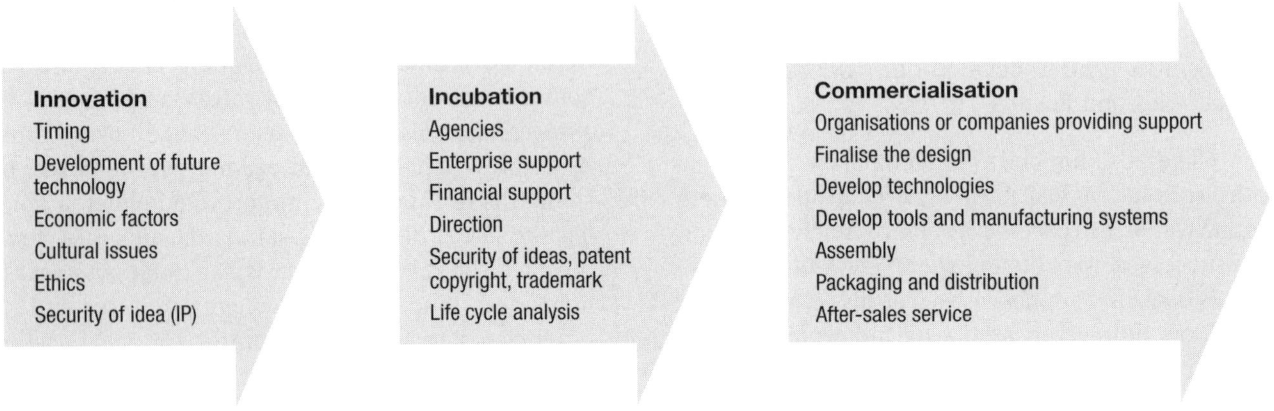

Innovation
Timing
Development of future technology
Economic factors
Cultural issues
Ethics
Security of idea (IP)

Incubation
Agencies
Enterprise support
Financial support
Direction
Security of ideas, patent copyright, trademark
Life cycle analysis

Commercialisation
Organisations or companies providing support
Finalise the design
Develop technologies
Develop tools and manufacturing systems
Assembly
Packaging and distribution
After-sales service

Figure 15.3 Entrepreneurial activity © R. Trimmer, 2010

15.2.4 Management and entrepreneurial activity

Developing an innovation to commercialisation requires management of a variety of different systems. The process occurs in three distinct stages as shown in Figure 15.3.

15.2.5 Legal and ethical issues

With the ever-increasing growth of entrepreneurialism and the consequent increase in the level of competition new problems are created by legal responsibilities and ethical behaviour. Entrepreneurs need to be proactive in cultivating effective legal risk management into their business venture.

Some of the social and ethical issues associated with computing and the Internet range from privacy to hacking, censorship to piracy of music and movies, social networking to computer crime. A continual revision of IP laws is necessary in these technological times. Individuals and companies are regulated by the government in the following areas of ethical standards.

- impact and quality of user-supplied web content
- privacy and computer technology: records of online activity, video surveillance, GPS location tracking, consumer dossiers, national ID systems, and so on
- Internet censorship laws and alternatives, spam, political campaigns regulation, anonymity and net neutrality
- IP—copyright laws, fair use, video sharing, software patents, free software, piracy and new business models
- computer crime—identity theft, hacking, credit card fraud, online scams, auction fraud, click fraud, stock fraud, digital forgery, and so on
- computers and work—job loss and creation, global outsourcing, telecommuting and employee monitoring
- errors, failures and risk—system failures, safety-critical applications, software design problems and techniques for improving reliability and safety
- how societies make decisions about new technologies.

Key concepts and definitions

Commercialisation—the commercial venture to transform innovations into economic viability.

Cradle to cradle design, sometimes abbreviated to 'C2C', is referred to as a regenerative approach to the designed articles where a proportion of the product is disassembled or recycled and reused in a way to avoid the ending of the life cycle and the ultimate consequences of landfill.

Entrepreneurial activity or entrepreneurship is the act of being an entrepreneur, a French word meaning one who undertakes an endeavour. Entrepreneurs assemble resources including innovations, finance and business acumen in an effort to transform innovations into economic gains.

Green chemistry, also known as sustainable chemistry, is the design of chemical products and processes that reduce or eliminate the use or generation of hazardous substances. Green chemistry applies across the design, manufacture and use of a chemical product.

Incubation—the development of an innovative design idea into an environment of developmental activity and observation in order to provide optimal conditions for heightened commercial success.

Innovation is often based on an invention that is modified or improved to suit a changing need.

Legislation—the act of making or enacting laws.

Litigation—the act of filing a lawsuit against an individual or a company. It generally develops from a complaint that requires resolution through a court of law.

Useful websites

- www.1000ventures.com/business_guide/crosscuttings/entrepreneur_main.html
- www.atp.com.au/
- www.ausicom.com
- www.ausindustry.gov.au/InnovationAustralia/Pages/InnovationAustralia.aspx
- www.ausinnovation.org/
- www.innovation.gov.au
- www.innovation.org.au/inventions
- www.indigenoustourism.australia.com/news.asp?sub=0638
- www.optalert.com
- www.virgin.com/about-us/
- www.workcover.nsw.gov.au

Classroom activities

1. Research the following entrepreneurs and list the indicators of their success.
   - Tom Potter: Eagle Boys Pizza
   - Dick Smith: Dick Smith Electronics
   - Naomi Simson: RedBalloon
   - Brian Singer and Doug Warbrick: Rip Curl Wetsuits
   - Fred DeLuca: Subway
   - Reg Mombassa: Mambo
2. List five agencies (government or semi-government) and in a T-chart describe their services.

Agency	Description of service

3. Identify a set of agencies that you could employ to help you bring your MDP to commercial success.
4. Developing an innovation to commercialisation requires management of a variety of different systems occurring in three distinct stages. With the knowledge you have received from researching your innovation case study, track the stages of management by placing them in the table below.

Case study		
Innovation	**Incubation**	**Commercialisation**

Key concept questions A pp. 211–212

1. What is meant by the term 'innovation'?
2. List four factors that impact on innovation and briefly explain how they do so.
3. Describe the work of agencies in bringing an innovation to commercialisation.
4. Name the four major design centres of the fashion world and locate them on a map.
5. Explain the role of an entrepreneur as they take an innovation through to commercial success.
6. List some of the qualities that make entrepreneurs so successful.

Sample HSC questions

Extended-response question A pp. 212–213

Rip Curl has designed the world's first power-heated wetsuit. Named the H-Bomb, it will change wetsuits forever by keeping users warm via heating elements that line the back of the suit. The heating elements are powered by two lithium ion batteries and are strategically placed to warm the wearer's core, which in turn allows the body to pump warm blood to the extremities. Rip Curl's H-Bomb is the future of cold water surfing.

1. (a) Rip Curl's H-Bomb is an example of a successful innovation designed specifically to meet a target market.
   - (i) Outline the main factor that has contributed to the success of the innovation. (2 marks)
   - (ii) Identify the functions that make the Rip Curl's H-Bomb different to existing wetsuits. (3 marks)

   (b) The key factor to any successful innovation is in the research and development. Outline the methods of research that would have been undertaken in perfecting Rip Curl's H-Bomb. (3 marks)

   (c) The protection of intellectual property is a vital asset for all designers.
   - (i) Explain why a protection strategy is necessary for Rip Curl. (2 marks)
   - (ii) List three strategies used to protect intellectual property and explain how they are used when designing. (5 marks)

16 – Creative designing and producing

HSC OUTCOMES

A student:	You learn about:	You learn to:
H3.2 uses creative and innovative approaches in designing and producing.	■ creativity and innovative design practice – processes undertaken to develop innovations – success of innovation – adaptation and development of ideas – responding to motivational stimuli – creative thinking.	■ demonstrate creativity in the development of the major design project. ■ critically analyse successful innovation. ■ discuss concepts of quality, innovation and creativity.
H6.1 justifies technological activities undertaken in the major design project through the study of industrial and commercial practices.	■ practices in industrial and commercial settings as they relate to the major design project including – safe work practices using selected resources – production techniques – selection of processes appropriate to an identified need or opportunity – collaborative designing and design teams.	■ identify design and production processes used in domestic, community, industrial and commercial settings. ■ implement safe work practices using selected materials and techniques in design and production of the major design project. ■ explain the principles underlying safe working practices and environments.

16.1 Creativity and innovative design practice

Many successful innovations are those that are surprising and new, whose concept shows originality of idea and thought. Innovation is not improvement: improvement is taking what you have and making it better. Innovation is about discovery: finding new materials, tools, manufacturing methods and making changes.

Innovators use the term 'thinking outside the box'. Designers 'stretch' their design thinking beyond the limitations of the situation. While designers need to be focused, the more inspired designers tend to shake things up by thinking ridiculous and far-fetched: in moving beyond the norm many designers find their true inspiration.

Student designers can be held back from developing a strong design sense. They may not place themselves into the context of designers and consequently not see themselves as such. Most design students have the potential to be innovative but they need to have the experiences that are the foundation of inspiration.

16.1.1 Processes undertaken to develop innovations

The processes in developing an innovation rely on the interaction between certain elements that influence the design activity. These elements follow a linear pathway involving a series of stages. Each stage needs to be completed prior to flowing on to the next stage. Stages may include:

- assessing existing and future needs of the consumer or user group
- researching existing ideas that might come close to meeting the needs
- designing a product, system or environment that will meet the identified need and legally register it as an innovation
- production, including industrial design, tooling, construction, assembly, packaging and warehousing of the innovation
- marketing, distribution and sales and servicing of the innovation.

Processes undertaken in the MDP

You can be innovative with your MDP. The first step is to take time to consider the idea and try to think 'outside the box'. Using lateral thinking or developing alternate ideas can be useful in allowing your ideas to grow. For more information on ways of thinking refer to Chapter 6.

The second step is to develop a strong vision of what you see as possible. The task of designing is much easier when you have a clear understanding of what is possible in the design.

The third step is to research existing designs. This is important to prevent overlap with other designers. An important government agency is IP Australia. This is the first stop in ensuring work is recognised as original and consequently supported by a patent. Patents give effective protection if you have invented new technology that will lead to a product, composition or process with significant long-term commercial gain.

The fourth step is to identify the market for the design. Using research data as early as possible will give you valuable information on the user group. This information will help in determining any established markets for the final design.

The fifth step is to design programmability into your innovation, providing a flexible nature to the product that allows it to be slightly redesigned at any required point. This also allows continued improvement to the product, building in design quality throughout the entire process of its development.

The sixth step is to design for 'fixability'. Innovations need to be designed so they can be fixed or repaired as this establishes a product's reputation for service reliability, which will reduce warranty costs and maintain a longer service life. There is an opportunity for designers to predict the associated ongoing repair concerns and also define product obsolescence.

Some designs fail at the product level. Many designers who have had a design failure would refer back to the production process to find the design fault. The design would then be improved using a system that included a trial and error application. Some innovations can still fail, particularly when the measure of success is profits or any other financial aspect. Failure here is permanent and cannot be rectified with trial and error design application.

Some of the greatest design 'faux pas' have occurred by a failure in the product design process when the linking of customer needs to product design and development does not occur. An excellent example is the original Hubble telescope, a billion dollar piece of space optics which, it was discovered after launch, had been built with the wrong focal length. The office gym exerciser and the gas-powered weed-cutting golf club are just two examples of the multitude of designs that did not follow the basics and failed in the product design to link with customer needs.

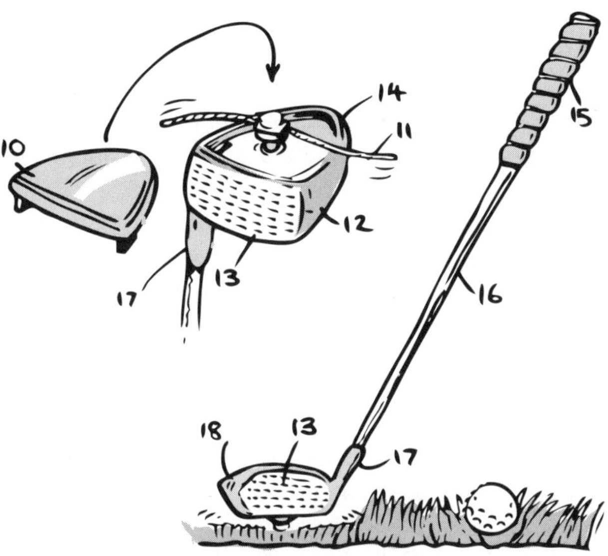

Figure 16.1 For use at home

Figure 16.2 For use in the office cubicle

Bad design, bad taste

The Ford LTD Family Truckster is iconic to many movie buffs. It was designed, modified and 'uglified' for the film *National Lampoon's Vacation* and never went into mass production. Its pea-green duco, spider-eye headlights and faux-wood panelling set new standards in bad taste. It sits among notable Hollywood cars such as the Batmobile and survived the film to now stand as an iconic emblem of bad design and popular culture. Sadly, Americans could buy mass-produced cars that did not look so different to the Truckster.

Figure 16.3 The Ford Truckster exploited all avenues of bad taste.

The American cars of the late 1970s and early 1980s were notorious for their styling excesses and the designers did not apologise for how unrelated styling elements were stacked unpleasantly atop each other.

Segway transportation device

Figure 16.4 In some designs it is hard to see the point.

Today's idea of 'innovate or die' betrays ignorance of product history. Plenty of innovative ideas have failed in the marketplace. For example, Segway's personal transport system has had success in niche markets such as warehouses and industrial sites but the Segway's capabilities compared to vehicles of similar price have limited Segway's market penetration. Also regulatory bodies have prevented usage due to the Segway's classification as a motor vehicle.

16.1.2 Success of innovation

True innovation introduces something new and/or re-invents a product, system or environment. Successful innovation is about ideas being applied positively in order to make improvements or increase productivity and wealth in the economy.

Executive managers like Steve Jobs at Apple, Yvon Chouinard at Patagonia and Sir Richard Branson of Virgin are good examples of leaders who promote the importance of innovation, who constantly stretch the boundaries and set goals for innovation. Their methods make innovation the expectation. While each company looks at innovation through its own eyes, there are striking similarities as outlined in the following:

- defining the tasks by taking risks and thinking creatively
- targeting existing as well as new customers
- considering the role and responsibility of the impact of the innovation
- defining the process and managing all the elements of innovation
- being goal driven and staying on track
- establishing collaborative innovation teams
- maintaining effective communication with others in the organisation
- maintaining continuous improvement via Total Quality Management
- sustaining investment in facilities, research and development and marketing.

16.1.3 Adaption and development of ideas

For designers to explore their creativity they often must work through a series of strategies to clarify their design thinking and assist in working towards innovative ideas. In other words, addressing an old problem in a new way and finding an alternative solution. Redesigning has positive implications for an object's obsolescence and can reduce company concerns over the product's service life.

Making vacuuming fun

The lack of efficiency in the old bag-design vacuum cleaners prompted James Dyson to develop a range of vacuum cleaners based on compressor technology that required no bag. The end result was a vacuum cleaner unique in its shape and using specific materials and industrial production processes. The design's functionality was a hit with householders and once commercialised, sales increased markedly. The transparent dust container was marketing genius as householders were able to look in and watch the vortex vacuum and the capture of dust. After a hundred years of stagnancy, the 'Dyson', with its focus on function and its innovation, challenged design norms.

New ways to use technology

Designers are constantly looking for and finding new ways to use technology which in turn allows the designer to explore a solution to a problem in a whole new way. Many innovations have been developed from technologies such as radio frequencies, microchip technology, carbon fibre materials, superconductivity, nanotechnology, infrared transmission and global positioning systems. Companies promote development through a variety of research and development operations. They use research and development as the best means of identifying and understanding how to use technologies in new ways. By doing this the company can become predictive in their planning. At this level companies also use systems to reduce the amount of risk and uncertainty involved in innovation development.

Air Guitar Pro

Virtual rock stars have become a sizable enough demographic for Takara Tomy to design the Air Guitar Pro, an air guitar player's tool that looks like a headstock with a tiny part of the upper neck and fret board outfitted with buttons that play chords. An infrared sensor embedded in the neck picks up strumming motions from the other hand and a built-in speaker provides amplification. This and many similar technologies are being used in digital gaming by companies such as Sony and Nintendo.

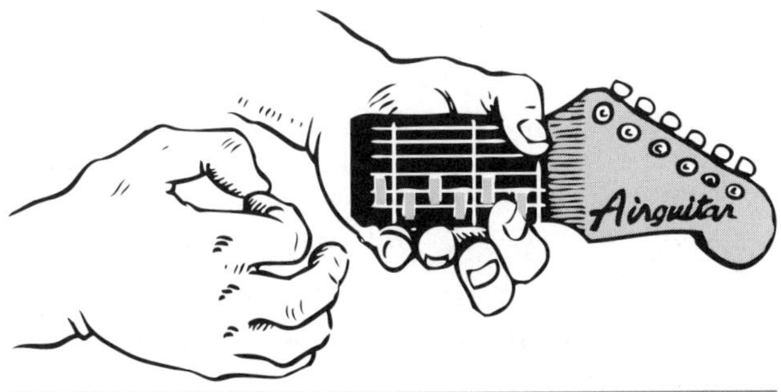

Figure 16.5 Air guitars now come in advanced models.

Sustainable design

Appropriate and ethical innovation needs to be environmentally sustainable. Sustainable design can be divided into five focus areas:

- avoiding waste
- reducing environmental and health impacts through eco-design
- reducing the environmental impacts through the development of educational curricula
- reducing the environmental impacts of production and consumption systems
- developing a life cycle assessment for the innovation.

A green milk jug for a 'green' earth

A major shift in the packaging of products has resulted in the rethinking of the plastic milk bottle.

Figure 16.6 Manufactured sustainability

It means that milk can now be stored and poured in a stylish container. The real advantage comes from the rethinking of the method of production: due to the new shape manufacturing and packaging costs have been reduced which has significantly reduced the environmental impact. The plastic used in the material has always been recyclable: the idea that a designer can choose a manufacturing process that reduces environmental impact is placing 'green' design on a new level.

Investigating new styles

Investigating new styles allows designers a new freedom to explore the aesthetics of design without too much compromise on the function. The style of design is viewed as creative, new and novel. The stylisation of an object in certain ways will evoke a feeling to the senses, and if successful will develop a meaningful connection between the designer and the end user.

Award-winning designer Marc Newson is innovative in the way he investigates new styles. Many of these are seen as interior designs in hotels in Europe. One of his more interesting designs was the interior of the Asrium spaceplane. His design is considered innovative in the way he has arranged the interior for passenger interaction, challenging the way seating is designed on modern aircraft.

Addressing new user groups

Addressing new user groups can widen the consumer base for design by increasing the market share. Some minority groups have particular needs that are not addressed by mainstream designers and to this end designers can modify existing products for different user groups. User groups are varied but may be defined by criteria such as physical and mental ability, disposable income or age. The usefulness of a product is measured through its level of support within the targeted user group.

Application of miniaturisation

Miniaturisation of larger products can move a design into a new environment. There is a growing trend amongst designers to explore miniaturisation and portability, and the public often reacts positively even when the object's usefulness might be compromised. The super pocket bike belongs to a whole new breed of motorcycle miniaturisation, being faster and more powerful than any predecessors. Ergonomically, this miniaturisation cannot go much further.

Figure 16.7 Small design, big fun, though miniaturisation has its limits.

Technology reduction

Designers use high-end technologies to make their designs work efficiently. All the technology we have at our disposal would be useless without suitable infrastructure such as electricity or infrared systems. There are parts of the world that do not have these supporting infrastructures and consequently designers are required to reduce the technology of existing designs. The task for the designer is to de-engineer a product to suit the technological level of the society. This concept is known by designers as 'downteching' and is commonly used in remote locations.

Professor Amy Smith and her team of engineers have been developing low-tech, low-cost technology for impoverished Peruvian villages. This has had the effect of fighting local poverty and is improving living standards in other developing countries where the technology is being used. Using the technology of the bicycle with its cranks, cogs and chains, engineers have been able to replace the power source of household machines to that of pedal power. Washing machines, water pumps for agriculture and battery-charging generators are all part of this established infrastructure. Apart from the benefits of redeploying the workforce, the technology is such that any part can be replaced or repaired by anyone with limited skill.

Professor Amy Smith's design theory works on what she calls the seven rules of low-tech design:

1. Try living for a week on $2 a day to understand what trade-offs need to be made with limited resources.
2. Listen to the right people to truly understand the problem.
3. Do the hard work needed to find a simple solution.
4. Create 'transparent' technologies: ones that are easily understood by the users and promote local innovation.
5. Make it inexpensive: when you are designing for people who are earning just one or two dollars a day you need to keep costs as low as possible.
6. To make something ten times cheaper, remove ninety per cent of the material.
7. Provide skills, not just finished technologies. Involving the community throughout the design process helps equip people to innovate and contribute to the evolution of the product.

Multifunctional products

Sometimes when a product includes a range of functions it can be considered innovative. Personal items such as mobile phones and watches continue to have more functions added. Although there is a design trend to increase function, there is also a level where designers will recognise that further increasing functions might render the original product useless.

The Swiss Army knife is iconic in the way it has encapsulated the concept of multifunction, having been developed, manufactured and sold over a 120 year period. The model with a USB drive has been around for a while although the latest improvement is increased storage capacity and some models have digital readouts for altitude and GPS plots. Swiss Army knives are synonymous with quality and multifunctionality and are recognised all over the world.

Figure 16.8 The Swiss Army knife: a classic example of a multifunctional design

16.1.4 Responding to motivational stimuli

Designers respond to external stimuli when gathering their ideas. Stimuli may include music, aromas, observations or photographs: almost anything in fact. For example, if designing a new computer mouse a designer might search for the right ergonomic shape by using a stone or modelling clay. The final outline may also be added to by a shape drawn from nature, such as a leaf.

Designers tap into cultural stimuli such as news and entertainment media and use them as a source of inspiration. Depending on the nature of the design, designers also respond to external conditions. Many designers seek ideas that can help global problems and address some of the critical issues facing today's world.

Engineers without Borders at Brown University are initiating a project to provide nearly a million people with clean drinking water by capturing rainfall in Kerala, India. The project will directly address a need in the community and work directly with local community members and organisations in design and operation of the water harvesting system.

16.1.5 Creative thinking

Creativity is a process involving the generation of new ideas or concepts. The designer discards any preconceived assumptions about a design and starts again from a completely new direction. At times the initial designs may seem ambiguous or multi-dimensional by their nature, but the process allows designers to pursue their creative ideas.

In 1899 Charles Duell, the director of the US Patent Office, suggested that the government close the office because everything that could be invented had been. Chester Carlson invented xerography in 1938. Virtually every major corporation, including IBM and Kodak, scoffed at his idea and turned him down. They claimed that since carbon paper was cheap and plentiful no-one would buy an expensive copier.

In 1995 the Volkswagon Motor Company launched a new model of the Polo car named the Harlequin. It was an unusual model featuring a multicoloured body with each panel in a different colour: red, yellow, pistachio green or blue. Instead of feeding the robot assembling the car with parts of the same colour it was fed with multicoloured parts. Initially meant as a joke, the idea gathered interest and sales increased for this model.

An important requirement for creative design is to be able to try out many different alternatives. Designers use a philosophy of 'high ceiling and wide walls' when designing, indicating the importance of a large 'design thinking space'. But in some ways creative design can go too far: when a design becomes 'over featured' with ideas this is referred to as 'creeping featurism'. This generally occurs when technology advances make it possible to add new features. A common belief among marketing experts is that designs are required to have more features to make them sell. The trend amongst designers is to offer simpler ways of doing things as reducing the number of features can actually improve the user experience. Most successful designers seek to understand their users in order to design products well-matched to their needs.

At times, creativity may not have an outward appearance. This is similar to technology's 'black box' concept where most of the technological components are hidden from the end user. The user interface or operational control is generally simple and streamlined. In creative designing much of this principle applies: designers work with creative flair, blending, changing and trying new ideas and the final design solution is a composition of creativity which to some degree is hidden from the user.

16.2 Practices in industrial and commercial settings as they relate to the major design project

When planning your MDP re-read Chapter 4 and then complete the following task. This task will form the information that you need to study for your possible exam question of comparing industrial and commercial settings.

Provide an image of your completed MDP and an image of one completed in either a cottage industry or by mechanised production. If possible, include all three. Label all diagrams with the functional and aesthetic properties of the products and compare these in a table.

16.2.1 Safe work practices using selected resources

Visit the NSW WorkCover website (www.workcover.nsw.gov.au/) and imagine you are applying for a job in an industry associated with your technology. List all of the OH&S aspects that you would need to consider should you be interviewed for a position.

16.2.2 Production techniques

For your own project and for a similar commercial project compare the design process, materials, tools, techniques and technology, method of manufacturing, and production sequence from plant layout to full-scale production.

16.2.3 Selection of processes appropriate to an identified need or opportunity

There are many production processes to consider when designing and producing. Justify the choice of production techniques used in the development of your MDP and predict how these techniques might be different if used in industry.

16.2.4 Collaborative designing and design teams

Identify the people who provided any assistance in developing your design solution and explain how they communicated with you about their role in the process. Compare this with research you have carried out on a commercial designer and their collaborations. See Chapter 1 for more information on this.

Key concepts and definitions

Design and production processes—the steps or procedures undertaken during the invention and manufacturing phases.

Domestic settings—homes where tasks are performed.

Faux pas—a slip in etiquette, manners or conduct; an embarrassing social blunder or indiscretion.

Industrial and commercial settings—any manufacturing, trade, cottage or production business.

Nanotechnology—the science and technology of building devices such as electronic circuits from single atoms and molecules.

Obsolescence—when something is in the process of passing out of use or usefulness.

Superconductivity—the phenomenon of almost perfect conductivity shown by certain substances at temperatures approaching absolute zero.

Technological activities—those tasks that involve using machinery or equipment during production.

Useful websites

- store.artlebedev.com
- www.dyson.com.au
- www.espdesign.org/sustainable-design-resources/sustainable-design-videos/
- www.intad.asn.au/conference/Russell%202.pdf
- www.segway.com/
- www.powerhousemuseum.com/exhibitions/success_innovation.asp

Classroom activities

1. Brainstorm the idea of an innovation with a small group. Once completed, share your results with another group and see which common elements the two groups share.
2. There are a series of processes designers work through to develop an innovation and some of these can also be adopted by students. List the processes undertaken in developing an innovative MDP.
3. We can learn much from innovative design failures. Research on the Internet and find a design failure. Describe the innovation and its faults and submit this along with an image of the design.
4. Companies recognise the importance of innovation for the long-term success of a product. Research a company such as Apple, Patagonia or Virgin that promotes the importance of innovation. Identify their strengths in the current marketplace.

Key concept questions A p. 213

1. For designers to explore their creative ideas they often must work through a series of strategies. Briefly list and describe each strategy for innovation.
2. How do designers respond to motivation and external stimuli when gathering their ideas?
3. An important requirement for creativity is to be able to try out many alternatives. What is meant by the analogy of designing with a 'high ceiling and wide walls'?
4. What is meant by the concept of the 'black box' design as it is related to the user?
5. List the methods used by companies to make innovation the expectation.

Sample HSC questions

Objective-response questions A p. 213

Circle A, B, C or D. (1 mark each)

1. Which of the following best describes an innovative product?
   - A a product with an effective marketing strategy
   - B a product that is used with a new user group
   - C a product that improves the level of technology
   - D a product that uses technology in a new way
2. The creative process in designing involves which of the following?
   - A experimenting and researching
   - B being able to try out many different alternatives
   - C working in a series of logical steps
   - D using skills and experiences to achieve a solution
3. The stimulus on the right depicts gym-styled equipment used to work out in an office cubicle. Unfortunately it was a design failure. The reason for this stemmed from its failure to
   - A locate cheap resources.
   - B develop a working prototype.
   - C link product design with customer needs.
   - D generate design ideas in the initial stages of design.

17—How to write a case study

HSC OUTCOMES

A student:	You learn about:	You learn to:
H6.2 critically assesses the emergence and impact of new technologies, and the factors affecting their development.	■ case studies on emerging technologies – factors affecting their development – criteria for evaluation – impact on society and the environment – impact on innovation.	■ appraise the ecological, economic, social, ethical and legal implications of new and emerging technologies. ■ analyse the impact of emerging technologies on innovation.

What is a case study?

A case study is an in-depth study of an innovation and the emerging technology that the innovation is built upon. An innovation involves a change to a product's design. There are three types of innovations:

1. a completely new idea or concept
2. an improvement to an existing design, and
3. a new design that has had a section of the design improved through problem-solving.

The innovation may involve a change to a product's design, manufacturing or marketing. Emerging technologies are new and/or improved materials, tools, techniques, products, systems or environments that allow innovations to occur. Innovations are usually linked to emerging technologies as people learn to use these in new ways. The main purpose of the study of innovation is to enable students to develop knowledge and understanding of examples of innovation and emerging technologies.

The Stage 6 NSW Design and Technology syllabus states that students will complete a case study of an innovation that includes reference to:

- design and design practice
- factors which may impact on successful innovation
- entrepreneurial activity
- the impact of emerging technologies
- the impact on Australian society
- historical and cultural influences
- ethical and environmental issues.

When selecting an innovation for your case study you must ensure that the information you gather will be enough to fulfil ALL of the requirements as if you miss a topic and the HSC question is on that topic, you may not be able to answer the question. Completing a case study is a time-consuming effort and you must be well organised if all areas are to be covered thoroughly. You are advised to gather all data before you start, dividing it up into each of the content areas. If you can find no information on any one area you will need to start again as a HSC examination question could be on any section of the content.

The resulting case study should be a useful, investigative final report that shows critical thinking and analysis using appropriate research techniques. The case study should have enough information to allow you to write an extended response on each of the syllabus topics. A good case study is more than just a description. It is information arranged in such a way that you become knowledgeable on the topic and able to discuss any aspect of it in depth.

There are three basic steps in case study writing: research, analysis and the writing of the report. You begin with research but even when you reach the report writing stage it is likely that you will need to go back and research more information.

The research phase

The secret to successful research is good organisation. You will need to find information on your innovation that relates to each of the topics including:

- the emerging technology
- the innovation
- designs and design practice
- factors which may impact on successful innovation
- entrepreneurial activity
- the impact of emerging technologies
- the impact on Australian society
- historical and cultural influences
- ethical and environmental issues
- creativity.

To ensure that you have covered each of the necessary areas you need to create a research folder. You could use an A4 folder divided into sections using A4 plastic envelopes that are clearly labelled with the above headings.

You can use secondary research via libraries and the Internet to explore what has already been written. The librarian will assist you to find a range of books, articles and other sources of information and you can use Internet search engines for further research. Use a range of information and remember to reference it as this will add validity to your research. Read and highlight the important points in each article and then write a summary. Place the article along with the summary (at the front) in the appropriately labelled plastic sleeve.

You will also use primary research by interviewing people who know about the innovation and/or emerging technology. You will need to track these people down and starting points include the Yellow and White Pages on the Internet. You may also know someone who could be a useful contact in this area of your research. You could interview the designer, manufacturer, seller or promoter or the consumer/user of the product in person, by telephone or by email. When you are interviewing people you must have a prepared list of questions to ensure that all necessary topics are fully covered. You will need a list of questions on each of the topics that you are researching. You will need to record their response either by taking notes or by recording them (with their permission). You also need to ask questions that will give you facts that might not be available from an article. Ask questions that do not let people answer with just a 'yes' or 'no' as you will get more information.

The analysis phase

Sort and organise the information into each plastic sleeve file. This may mean cutting up and dividing an article into each of the headings. Once you have collected your information from people, articles and books you need to think about it, sort through it, take out the excess and arrange it so that you can write an in-depth summary on the topic. The summary must include data on what the reader would need to know in order to understand the topic.

Writing the case study report

The report must demonstrate higher-order and critical thinking, evaluation and synthesis using appropriate research methods. The case study report must include all of the following elements.

Title page

The title page is the cover sheet and contains the name of the case study and the author.

Contents page

The contents page is a list of the section headings. It may look something like the one opposite.

Abstract

The abstract is a 100–150 word summary on the innovation and emerging technology with a brief description of how the two are interrelated. This should be accompanied by a list of all of the topics that are discussed in the main body of the report.

Acknowledgements

The acknowledgements are a list of the people and/or companies who have assisted you. These may include family members, teachers, a representative from a company, the people you interviewed and anyone who supported your work.

Main body of the report

This is the detailed section of the report. The whole purpose of writing case studies and sharing them with others is to share the experience without everyone having to do the research.

You will need to organise the main body of the report. You can include an introduction describing the innovation and emerging technologies and discussing how they are linked together, some background and historical data on the innovation's development and then further detailed information on each of the content areas.

The content areas include every heading listed on the contents page in the sample table above and will consist of a summary of every aspect uncovered in your research. You may find at this point that additional research is required in order to cover a section thoroughly. It is suggested that you address the issues listed under the following headings in order to gather sufficient data.

Emerging technology and innovation

- Name, describe and provide a labelled graphic of the innovation and the emerging technology.
- Explain why this design is innovative.
- Name the designer and the design company manufacturing and marketing the product/s.

Designs and design practice

- Name the designer and describe their role in the company (if applicable).
- Find out when and where this design was created.
- Find out why it was invented (what was the need) and if there were any limiting factors.
- List the steps in the process used to design this product.
- Describe the research that was carried out before finding the design solution.
- Explain how the ideas were generated or developed.
- Discuss the factors that limited or contributed to the success of the innovation.
- List any alternate solutions the designer considered.
- Evaluate the functional and aesthetic qualities of the design.
- Explain the role of the new or emerging technology in this design.
- Identify the factors that contributed to the quality of the design.
- Analyse the durability of the design.
- Justify which experiments would be most useful when developing this design.
- List the resources that were used in the development of the innovation.
- Describe the problems or limitations of the design.
- Find out if ergonomics were considered in the design and if so, explain why this was important.
- Evaluate the importance and role of ongoing evaluation in this process.
- Evaluate how well the solution fits the problem.

Factors which may impact on successful innovation

- Discuss the timing of the innovation.
- Describe how available technologies impacted on the development of the innovation.
- Analyse the cultural factors that impacted on the development of the design.
- Identify and evaluate the extent to which political, economic and legal factors impacted on the innovation.
- Discuss the demand for the innovation and identify the market group.
- Suggest a marketing plan for the innovation.
- List the agencies used in the development of the innovation.

Entrepreneurial activity

- Describe the impact of the entrepreneurial activity on the success of the innovation.
- Describe the management structure and its role in supporting the innovation.
- Evaluate the legal and ethical issues considered by the entrepreneur.
- Explain how government, commercial and industrial agencies have impacted on the innovation.

The impact of emerging technologies

- Analyse the impact of continually emerging technologies on this and future innovations.

The impact on Australian society

- Explain the impact of this innovation on Australian society including the social, global, political, economic, environmental, technological, legal, cultural and marketing spheres.

Historical and cultural influences

- Timeline the development of the innovation.
- Add to the timeline all factors that have either contributed to or hindered the innovation's success.
- Describe the impact of Australia's social diversity on the innovation.
- Identify the changing social trends that may impact on the innovation.
- Explain how the changing nature of technology and work has impacted on the innovation.

Ethical and environmental issues

- Discuss each ethical issue associated with the innovation.
- Explain how the designer protected their intellectual property.
- Describe how issues such as copyright and plagiarism impact on the designer.
- Discuss the short- and long-term environmental consequences of the design.
- Explain how sustainable technologies were considered in the design.
- Describe how obsolescence may be a part of the innovation.
- Evaluate the product in terms of its ecological footprint.
- Create a life cycle analysis of the product. Summarise your findings.
- Discuss the rights and responsibilities of the designer from an environmental perspective.

Creativity

- Identify every area where creativity was used in the design or the design process.
- List the creative aspects of the design solution.
- Use a flowchart to demonstrate the manufacturing process of the design.
- List all of the cognitive organisers used by the designer.
- Discuss what motivational stimuli and creative processes were used by the designer.

Every heading must be covered in depth within the body of the report. This is important because there may be questions in your HSC examination on any of the listed topics. If you miss a topic, and it is in the exam, you have disadvantaged yourself.

Conclusion

Your case study will need a conclusion. To do this you must write a simple paragraph or two that summarises the main points of the case study. Once you have completed the report it is important to keep researching and updating your information right up until your final exam.

Reference list

Every reference book, DVD or Internet site used must be detailed in the reference list. There are a number of different referencing styles available for use. Your teacher may nominate a preferred style; if not, the following is one acceptable version.

Texts and books

Author surname, then initial, year published, name of book, publisher and place of publication. For example, Trevallion, D., Trimmer, R., *Excel Preliminary and HSC Design and Technology*, Pascal Press, Sydney, 2010.

Journals

Author surname, then initial, year published, name of article, name of journal, volume (issue number), pages. For example, Middleton, H., 'Creative thinking, values and Design and Technology education', *The International Journal of Technology and Design Education*, 1(15), 61–71, 2005.

Websites

Retrieved on 18/2/09 from [insert website URL].

Note that when using a number of books or journals they are listed alphabetically by the author's surname. Indentation always occurs on every line after the first one. URLs are always listed alphabetically from the first letter after www.

Appendix

Your appendix consists of any material that supports the case study and has been referred to in the written report. The information is grouped according to content and labelled as Appendix A, B, C, and so on.

List of figures or details

If you have referred to any tables, graphs or grids in your report you place a labelled copy of them here.

Tips for a completing a successful case study

- Know the innovation in-depth before you begin your case study analysis.
- Give yourself enough time to write the case study analysis. You do not want to rush through it.
- Be honest in your evaluations. Do not let personal issues and opinions cloud your judgement.
- Be analytical, not descriptive.
- Proofread your work!
- Ensure you have covered every heading from every perspective.

Key concepts and definitions

Case study—an in-depth study of a particular topic. In Design and Technology the topic case study is to be carried out on an innovation and the emerging technology that the innovation is built on.

Primary research—using first-hand information that you have collected. This is research gathered from people using surveys, questionnaires, interviews, and so on.

Secondary research—using information that someone else has gathered. This can be obtained from existing sources such as the Internet, books, magazines, videos, journals, and so on.

Useful websites

- business-project-management.suite101.com/article.cfm/steps_to_analysing_a_case_study
- www.bell.uts.edu.au/awg/case_studies/analysing
- www.boardofstudies.com.au
- www.northwestern.edu/african-studies/ugandaresearch/submit/howtowrite.pdf\
- www2.wmin.ac.uk/haberba/caseanal.htm

Classroom activities

1. Research and suggest alternate formats and headings that could be used when writing a case study.
2. List five innovations that you may choose to write your case study on and list five URLs related to each innovation. The URLs should ideally provide a range of information on each of the headings.
3. Check that the information you have gathered on each of the five innovations covers all of the questions that may be asked in the HSC.

Key concept questions A p. 214

1. List the labels that you will place on each topic sleeve in your case study folder.
2. How will you know if you have enough information to complete your case study?
3. Once you have your information summaries, what headings will you use in your information report?
4. What are your options if part way through the report you find that you do not have enough information on one of the topics?
5. Describe the difference between an innovation and an emerging technology.

18 — Innovation and emerging technologies: Case study 1—Optalert

HSC OUTCOMES

A student:	You learn about:	You learn to:
H6.2 critically assesses the emergence and impact of new technologies, and the factors affecting their development.	■ a case study on emerging technologies including – factors affecting their development – criteria for evaluation – impact on society and the environment – impact on innovation.	■ appraise the ecological, economic, social, ethical, and legal implications of new and emerging technologies ■ analyse the impact of emerging technologies on innovation.

18.1 Case study on emerging technologies

18.1.1 Factors affecting their development

The need

The Australian Transport Safety Bureau estimates that driver tiredness results in thirty per cent of all fatal crashes. The greatest danger with driver drowsiness is that drivers are not aware when they are too drowsy to drive. Research has shown that drowsiness impairs our abilities and awareness of situations and until now there has been no way of accurately determining how drowsy a driver is at any particular time.

There is much current advertising warning drivers of the danger of 'micro sleep' and of the need to 'Stop, Revive, Survive'. Television advertisements featuring Dr Karl Kruszelnicki have reinforced this message.

18.1.2 Criteria for evaluation

Driving while drowsy is one of the main causes of fatal road accidents worldwide. It requires only a few seconds to drive off the road and collide with a roadside object or another vehicle. Drowsy crashes often occur at full speed, with no evasive action taken. It is now recognised that a drowsy driver who causes an accident is as legally culpable as a drunk driver. Culpability can now extend to transport industry managers.

Figure 18.1 Truck drivers are at risk of fatigue when travelling long distances.

Professional drivers, such as long-distance truck drivers, do not have the luxury of stopping and resting as they usually have tight deadlines to meet. Health and safety, and particularly drowsiness, is a primary concern for both transport and logistics companies and individual drivers. Customers, insurance companies, governments and regulators are placing demands on transport companies, vehicle manufacturers and individuals to address the issue of accidents caused by drowsiness.

The innovation

A much needed innovation was some form of unit that could identify the state of drowsiness in a driver and warn them in some way. A new technology called Optalert uses eyeglass frames fitted with tiny sensors that help alert motorists when their fatigue reaches a dangerous level. The trouble signs of fatigue such as excessive blinking or rapid facial movement around the eyes alert the sensors, resulting in a high-pitched audio warning.

Figure 18.2 Optalert detects driver fatigue.

This design combined two existing technologies to form a new innovation. The first was to determine the drowsiness factor that would be used to test and monitor fatigue. The second technology involved the physical development of the technology for use in glasses and sunglasses. The designer used a monitoring beam to detect fatigue and the idea of supporting this in an eyeglasses frame completed the innovation. The only research and testing that had to be done was to develop a system able to identify and process the information recorded.

Optalert technology

Sensors are held in the glasses and detect low levels of infrared light to sense movements in the eyes and eyelids. By calculating the drowsiness scale every minute, Optalert gives a measure of a driver's drowsiness at any point in time and provides predictive warnings before the drowsiness reaches a dangerous level. The information can be recorded and returned to a remote location. This could be easily developed forward to allow companies to monitor their drivers and fleet.

Research and development

The company which developed Optalert, Sleep Diagnostics Melbourne, is an Australian business founded by Dr Murray Johns. Through his work he could see the potential risks of people falling asleep at their jobs. The most identifiable risk occupation Dr Johns identified was long-distance truck driving. During this time he developed a scale of drowsiness recognising the dangerous 'micro sleep'.

Finance was obtained and research commenced, with Dr Johns using theories based on thirty years working with sleep disorders and sixty-five published papers. Dr Johns was able to secure intellectual property for his idea, providing him with the security that no one could steal the concept and become a competitor. He then set out to manufacture the components by engaging the support of an innovation manufacturing company, Compumedics Limited.

Further innovation is likely, with BHP Billiton and other companies researching Optalert in their mines and additional research being carried out on the technology. There has also been interest in its application by the military, who are independently conducting their own trials. The technology has been developed for quite some time but there will need to be some final decisions made before commercialisation is sought.

Design development problems

The design team had to synthesise a sensing device to recognise the facial features. They first looked at previous devices used in sleep laboratories and hospitals and decided that these methods were impractical. The team worked on the PERCLOS method where video was used to measure eyelid closure. It looked successful in the early stages but the problem the team encountered was that the video monitors had difficulty capturing images reliably when driving in sunlight with shadows. The design team was also concerned with the 'human factor', where drivers might rely too heavily on the technology when wearing Optalert. As drivers could think that 'nothing can go wrong now', marketing needs to reinforce the idea that Optalert does not replace humans identifying their own tiredness.

18.1.3 Impact on society and the environment

Environmental impact

Currently companies are required by law to develop and maintain environmental standards specific to their technology. Many companies now see the 'green design' tag as both good business and responsible design. It also makes good business sense for the company to develop clean and sustainable manufacturing technologies from the start. For commercial applications, the Optalert technology will be incorporated into conventionally-styled plastic sunglasses. The idea of reuse is an important feature in sustainable management. There is also allowable design for the Optalert technology to refit or retrofit into prescription glasses or sunglasses. Optalert has commenced research into materials but at present the prototype has been derived from plastic found as a non-sustainable resource. The company sees that this will change shortly as the product moves to sustainable work practices involving recycled plastics and the use of inter-replaceable parts.

Social impact

Trends in society show that the demands of work and leisure leave a significant proportion of people chronically deprived of sleep. This can have damaging effects on productivity, health and safety. As this current trend of sleeplessness is maintained it will continue to cause a significant proportion of road and other accidents. One example is the Exxon Valdez oil tanker, which led to one of the worst environmental catastrophes in history. With tight profit margins and deadlines, drivers are putting their lives and those of others at risk by staying awake to earn a living. As the distance between home and work increases in most metropolitan cities, the problem of drowsy driving will worsen.

Marketing

People become potential customers after an enlistment campaign involving marketing. As the product is a 'one-off' purchase for most people the company may not be relying on too many repeat sales. This may be different, however, when it comes to large transport companies requiring the technology replaced over a period of time. For continued sales to the same people Sleep Diagnostics will need to establish strong commercial relationships with its customers. As glasses are also a fashion statement to many people, a range of styles of Optalert glasses will be developed.

18.1.4 Impact on innovation

Success of the innovation

The success of many innovations seems to be linked to profit. The product makes its designers rich and the innovation becomes a household name. But there are other measures of success. Many designers begin a design brief with the idea of making the world a better place, seeking satisfaction beyond the wealth factor. They see their innovation saving lives and making a difference, as was the case for Dr Johns. Dr Johns succeeded because, like many of the best and most successful designers, he used his vision as his driving force and the basis of establishing his criteria for success.

New product development agencies and organisations

A variety of agencies provided support for Sleep Diagnostics at appropriate times, working alongside them much like a team. They included the following.

- financial support: NRMA. The NRMA Motoring & Services invested up to one million dollars in Optalert. The motoring organisation saw this as an opportunity to save lives and also reduce insurance costs.
- programmed support: AusIndustry is the Australian Government's agency for developing and supporting Australian innovation. It does this through a variety of support programs: Optalert was considered by COMET (Commercialising Emerging Technologies) and this was an opportunity to commercialise the Optalert innovation into reality.
- manufacturing and outsourcing: Sleep Diagnostics needed to outsource the manufacturing for Optalert and found support in the company Startronics, an electronic manufacturing company providing specialised services to 'starting out' companies. As part of its support it provides testing, manufacturing, assembly, packaging and distribution of the final product.

Publications related to Optalert

Johns, M., Tucker, A., Chapman, R., 'Monitoring the drowsiness of drivers: a new method based on the velocity of eyelid movements', delivered to International ITS Conference, San Francisco, November 2005.

Johns, M., Tucker, A., Chapman, R., 'A new method for monitoring the drowsiness of drivers', delivered to International Conference on Fatigue Management in Transportation Operations, Seattle, September 2005.

Johns, M., Tucker, A., 'The amplitude-velocity ratios for eyelid movements during blinks: changes with drowsiness', delivered to Associated Professional Sleep Societies Annual Scientific Meeting, Denver, June 2005.

Johns, M., 'The amplitude-velocity ratio of blinks: a new method for monitoring drowsiness', *Sleep*, 2003; 26 (Suppl): A51-52.

The Australian, 'Local sleepiness alarm first to market', Optalert featured in the motoring section on 8 June 2005.

The Courier Mail, 'Specs Keep Drivers Safe', reporter Joel Dullroy profiles Optalert, October 2005.

Key concepts and definitions

Criteria for evaluation—a list of factors that determines whether the solution to the problem has been successful. Each item is evaluated to determine this.

Emerging technologies—those that are currently being developed.

Impact on society and the environment—the influence of the design solution on humanity and their surroundings, including the natural world.

Innovation—the creative solution to a problem. Innovation often uses an emerging technology to result in an original resolution. The innovation may be a product, system or environment.

Retrofit—the measures taken in the manufacturing industry to allow new or updated parts to be fitted to old or outdated assemblies.

Useful websites

- www.ausindustry.gov.au
- www.facebook.com/topic.php?uid=139000253585&topic=10703
- www.monash.edu.au/muarc/about/MUARC%20overview.pdf
- www.onepetro.org/mslib/servlet/onepetropreview?id=ASSE-02-513&soc=ASSE
- www.optalert.com
- www.patentsonline.com.au
- www.rta.nsw.gov.au/roadsafety/fatigue/stoprevivesurvive.html

Classroom activities

1. Develop a list of the variety of agencies that provided support for the company Sleep Diagnostics. Indicate in what ways this support was provided.
2. Brainstorm how this innovation may be used in other areas and technologies.
3. Identify the problems due to accidents caused by drowsy driving. Research how government initiatives are working to reduce these problems.

Key concept questions A p. 214

1. Name and describe the innovation in this case study. List the practices used by these designers.
2. Identify and describe the factors that made the innovation a success and the factors that impacted on its development.
3. Explain how entrepreneurial activity was used to promote the innovation.
4. Describe the emerging technology that contributed to this innovation and suggest how further development may enhance the product.

Sample HSC questions

Short-answer questions A p. 215

1. Analyse the impact of your innovation case study on Australian society. (3 marks)
2. Draw a timeline that traces the historical development of the innovation case study. (2 marks)
3. Describe the impact of cultural influences on the innovation case study you have chosen. (2 marks)
4. Outline the social and ethical issues associated with your innovation case study. (3 marks)
5. Evaluate the environmental impact of your innovation case study on society. (5 marks)

19 — Innovation and emerging technologies: Case study 2—Fast swimsuits

HSC OUTCOMES

A student:	You learn about:	You learn to:
H6.2 critically assesses the emergence and impact of new technologies, and the factors affecting their development.	■ a case study on emerging technologies including – factors affecting their development – criteria for evaluation – impact on society and the environment – impact on innovation.	■ appraise the ecological, economic, social, ethical, and legal implications of new and emerging technologies. ■ analyse the impact of emerging technologies on innovation.

19.1 Case study on emerging technologies

Figure 19.1 Range of fast suits

19.1.1 Factors affecting their development

The need

To win an Olympic gold medal is a lifelong dream for many athletes and the culmination of years of training and dedication. In Australia, to win Olympic gold is to become a celebrity, a superstar, an Australian hero. The sponsorships that accompany the glory of gold can provide financial stability so the importance of wearing the best possible swimsuit, including the innovative fast suit, cannot be underestimated.

Competitors use every advantage available to them including the best coaches, the latest training techniques, the most efficient swimming styles, the greatest fitness levels and the best equipment available, including fast suits.

All competitors must decide if they need one to be competitive. This will also be determined by the amount of money available, as the cost is approximately $500 per suit and they can only be worn about five times before they stretch and thereby lose the competitive advantage.

Background information

What works against the speed of a swimmer? The fluid forces in water slow swimmers down. The strongest force pushing the swimmer backward is pressure drag. Drag acts stronger on the surface than underwater. As a swimmer pushes on the water, the water pushes back and, as they swim, they displace water which travels as a wave in front of them. This, known as wave drag, adds to the pressure pushing the swimmer backward.

A weaker force is skin friction drag, where a sheet of water sticks to the swimmer's body and moves with them. Farther away, still water pulls on the water attached to swimmer, further slowing them down. As the swimmer and the sheet of water pulls away from the still water eddies are created. These eddies travel a few centimetres around the swimmer in what is called the boundary layer and this slows the swimmer down even more.

Friction drag worsens with the amount of skin in contact with the water. If the skin surface is rough, this slows the person down as it is harder for water to flow between their body hairs. The amount of drag can be controlled, to a degree, as below.

- Lowering pressure drag by the swimmer controlling how they present themselves to the water.
- Reducing wave drag from the dive. When swimmers dive off a block they spend a long time underwater, thereby reducing surface drag. If they float higher in the water when they emerge from the dive they will push a smaller wave in front of them, further reducing drag.
- Shaving body hair cuts drag by smoothing the surface. The less roughness, the easier the water will flow over a person.
- Strong abdominal muscles help swimmers stay straighter and float higher in the water. With a strong upper body there are fewer problems with drag because of the compensation of more powerful strokes.

If a swimmer's technique and body are in top form and they still want to swim faster, then 'someone' wants to sell them a fast suit, an innovative swimsuit using the latest technology.

Emerging technology

The latest technologies surrounding racing swimsuits are aimed at:

- reducing drag in the water
- creating fabrics that cling to the body and repel water

- creating fabrics that compress muscles and mould the shape of the body
- bonding fabric seams together while still allowing stretch
- replicating a shark's dermal denticles to reduce drag and increase speed.

Reducing drag

The emerging technology is a fabric that is extremely lightweight: a high density microfibre woven from chlorine resistant elastane and ultrafine nylon yarn.

Speedo, who are a leader in the development of this technology, explain that this 'Fastskin' allows for various water flow patterns over the whole body, providing a smoother swim and reducing drag by up to four per cent.

Water repellent fabrics

The fibres are made from nylon which is a hydrophobic (water-hating) molecule. This means that very little water enters the fibres. The fibres are manufactured into fabric and a finishing process applied. The finishing is a bonding process that produces a durable, long-lasting water-repellent finish. This finish was developed for use by Speedo in 2006.

Muscle compression

The fast suit is made from 'super stretch' fabric to improve the fit while compressing muscles. It is laser pulse, which means it is an extremely lightweight and water-resistant fabric and only one to two millimetres thick. It has very low skin friction and the result is a reduction of drag and muscle vibration.

As an additional feature, in 2006 Speedo incorporated gripper fabric in the Fastskin along the forearm area. The gripper fabric is used to mimic skin and replace the swimmer's sensitivity, increasing the feel and grip of the water through stronger friction. This friction is required for the swimmer to pull themselves through the water.

Bonded/welded fabric seams

In 2008 Speedo engineered bonded and then welded seams to follow the direction of the water flow and reduce drag. Comfort and freedom of movement is created by the incorporation of a super stretch thread. Speedo created an anatomic/dynamic pattern where seams act like tendons and provide tension in the suit while the fabric panels act like muscles, stretching and returning to their original shape.

Replicating dermal denticles

Due to its speed in water the shark was used as a model for the swimsuits. The shark's pace is attributed to V-shaped ridges on its skin called dermal denticles, which decrease drag and turbulence around its body, allowing the shark to move through the water more smoothly. This feature has been introduced into the swimsuit.

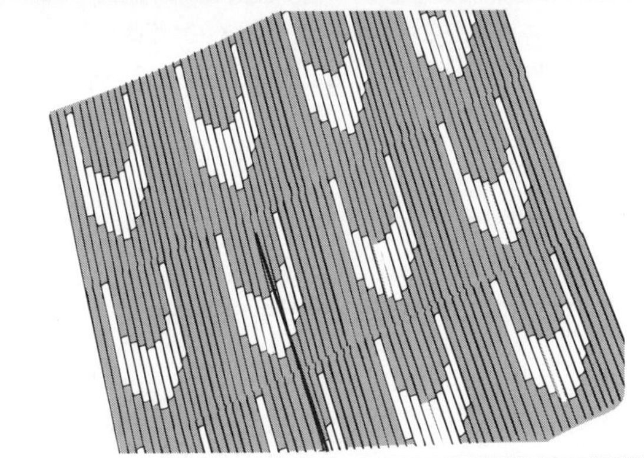

Figure 19.2 Dermal denticles emulate the skin of a shark.

19.1.2 Criteria for evaluation

Below are the major evaluation criteria for the swimsuit.

- It must not add drag.
- It must reduce drag to assist faster swimming.
- It should support the swimmer's conservation of energy and/or oxygen.

Development of the Fastsuit

In 1996 Speedo decided to attempt to produce the fastest swimsuits possible by improving on their Aquablade swimsuit. In attempting this they created the Fastskin. In 2004 they introduced the Fastskin II (FS-Pro) which reduced the total amount of drag over the suit's surface and was a move away from the traditional style of swimwear.

Speedo's focus on managing existing forces and matching them to make best use of an athlete's individual strengths resulted in the development of a number of suits: the Fastskin FS-Pro, the FSII and the LZR Racer.

In 2000 FINA approved the first full-body swimsuits that claimed to reduce drag. Swimmers sported various body and leg suits at the 2000 and 2004 Olympics, culminating with the debut of Speedo's LZR Racer in 2008. This decision has recently been reversed though swimmers such as Michael Phelps will retain their records and medals.

FS-Pro	FSII	LZR Racer
■ Preferred by Michael Phelps ■ Lightweight water repellent LXR pulse fabric (patent pending) ■ 3-D suit creates optimum performance ■ 15% more power and compression, increased stability, power transfer and reduced muscle oscillation	■ Preferred by many elite athletes ■ Is the 'shark suit' worn by elite athletes ■ Super stretch seams ■ Uses biofuse, a process whereby hard, skeletal properties are fused with flexible, tactile materials ■ Uses dermal denticles	■ Developed by NASA ■ Smoother than skin due to structure of fabric, like kitchen cling wrap ■ Three pieces are welded together ■ Welding reduces drag by 6% ■ The tight fit of the suit compresses a swimmer's muscles, adding to smoothness ■ Tightness affects breathing patterns so swimmers must train in the suit ■ The stabiliser, a girdle in the middle, allows easier floatation and balance which is useful in longer races when muscles tire ■ Suit is very light ■ In simulations swimmers are 4% faster and use 5% less oxygen ■ Allows a range of body movement and is energy efficient

Design research and development

Aqualab is Speedo's US research and development labarotory. Here the design team continually push the boundaries of sport science and technological innovation to create leading edge swimsuits. To achieve this, the teams work with experts from diverse industries including aerospace, engineering, medicine, sports science, textile technology and garment construction.

The resulting innovations are rigorously tested with partner athletes. For example, the Fastskin FS-Pro was designed using data provided from the body scans of hundreds of elite swimmers. In order for Speedo to virtually analyse the drag and flow of water around swimmers, testing and analysis included a highly sophisticated process called Computational Fluid Dynamics (CFD) from which they could create a virtual swimmer.

19.1.3 Impact on society and the environment

What does any society or a country like Australia have to gain from swimming competitions? It is generally accepted that athletics has a positive impact on society: physically, recreationally, emotionally and psychologically. The philosophy behind modern athletics is that an individual can achieve on his or her own. The line between acceptable and unacceptable 'help' has been blurred by some athletes using illegal performance-enhancing drugs. Competitive swimwear falls under the 'acceptable' category at the moment but it is interesting to imagine under what circumstances competitive swimwear could become 'unacceptable'. Is there or should there be a limit to how far advances in athletic equipment should go?

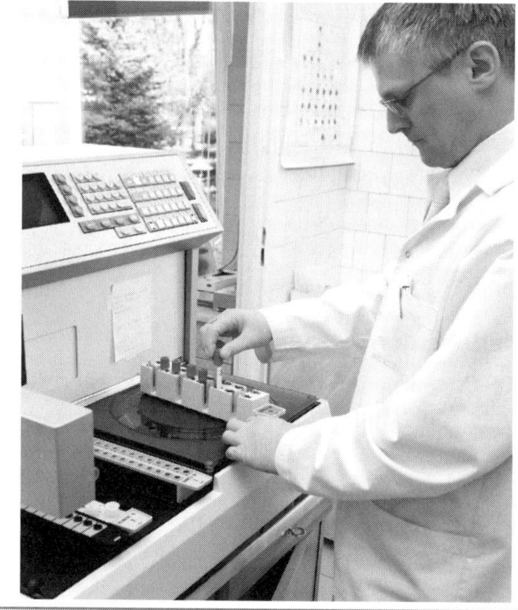

Figure 19.3 Research and development of fast suits begins in a laboratory.

Ethical issues

Many people see fast swimsuits as performance enhancers. The opposing view is that the suits do not enhance the swimmer's performance beyond their ability but simply allow them to reach an optimum level without the interference of a poorly manufactured swimsuit. Supporters would argue that this is just a part of the progression in technical improvements that has been going on for nearly one hundred years.

Other people view this evolution in technology as a disadvantage for poorer countries that cannot afford to adopt the new range for their swimmers. Many feel this removes the 'even playing ground' for all athletes and leads to more philosophical questions related to how much technology should be used to assist an athlete.

The following is a summary of communications received from scientists who are investigating the properties of new swimming equipment. In Australia, prominent aquatics scientists and researchers have found that if the suits are worn dry they are buoyant and assist the swimmer. When bodysuits are wet, there is no buoyancy (floatation) assistance. A floatation test found that a fastsuit takes approximately seven minutes to sink, proving its buoyancy.

In Europe, scientists and researchers claim that suits increase buoyancy and affect muscle movement so the suits should give physiological and biomechanical benefits. It takes time to learn how to use the suit, as with any other new equipment. With multi-conglomerates spending millions in developing fast suits, those who oppose their introduction are losing the battle.

Environmental issues

The fast swimsuits are custom-made for each elite athlete, meaning that they cannot be mass- produced. This means that environmental issues such as manufacturing pollution and extreme wastage are minimised.

Swimmers are wearing complete bodysuits requiring more fabric which in turn results in increased costs. Each panel is cut and moulded specifically according to the tested pattern, thus increasing waste. The fabric is made from nylon and elastomeric fibres which in turn are derived from oil, a non-renewable and depleted fossil fuel. This is not an example of responsible or sustainable design as the carbon footprint that these swimsuits leave is notable. Obsolescence is also an issue as the suits are easily torn and stretched and cannot be repaired.

19.1.4 Impact on innovation

Some of the various impacts on innovation are outlined below. Impacts on the consumer include the following.

- The suits are proven performance enhancers for swimmers.
- They are water resistant, reducing drag and muscle fatigue by up to four per cent.
- They are lightweight for a firmer fit.

Impacts on the manufacturer include the following.

- All three suits have proven to be extremely successful, with over forty world records broken by wearers.
- Due to the obsolescence factor there is constant demand and continual manufacturing.

Impacts on the environment include the following.

- The suits are not mass produced so this will not impact on issues such as manufacturing pollution or extreme wastage.

Factors which hinder successful innovation

Factors which hinder successful innovation are outlined below. For consumers these include the following.

- Fast swimsuits are expensive, ranging from $400–$600.
- They are not available for purchase by the general public but readily available to elite athletes.
- They take from twenty to thirty minutes to put on.
- The ultra-thin, lightweight, super stretch fabric means the suits can easily tear.
- Suits stretch after several wears and become ineffective.

For manufacturers these include the following.

- Due to the specific design pattern of suit panels there is an increase in fabric waste making the suits costly to produce, hence the high purchase price.
- Controversy over the issue of performance-enhancing suits may taint the manufacturer's name and reputation.

For the environment these include the following.

- Dramatic fabric waste due to each panel being cut and moulded specifically to the athlete's measurements.
- The suits quickly become obsolete due to wear and tear.

Figure 19.4 Fast suits are still not readily available to the general consumer.

Entrepreneurial activity

The competitive swimwear industry continues to grow because the market for the 'ideal swimsuit' is large. Swimmers use different strokes and value different designs that allow more flexibility according to their chosen swimming style. However, for all swimmers there are several key components that combine to create the ideal swimsuit.

- A low weight keeps the suit from absorbing water: it should be a lightweight material and it should be water resistant rather than absorbent.
- Reduced drag via a super-secure fit reduces skin friction and the drag caused by the roughness of the surface of the object. The suit should be hydrophobic with a polyamide (plastic) coating, woven rather than knitted to prevent stretch, and ultra-thin.
- The suit must provide physiological (physical) comfort through assisting optimal blood circulation to slow lactic acid build-up and prevent muscle soreness by reducing muscle vibration and providing full mobility and flexibility at joints.
- Materials should produce the least skin drag and have a polyamide coating that is hydrophobic, creating more glide and less friction. It must be elastic so that it is very form-fitting. It must be made of nylon which is resistant to breakdown from chlorine in pools.
- The shape and form should be flexible at joints and tight fitting everywhere else to allow circulation.
- In regards to aesthetics, the shape and form will be influenced by style, fads and fashion as well as functionality. The suits must be of a pattern or colour appropriate for serious competition.

Almost all swimwear companies including Speedo, Nike, Adidas and Arena make a fast suit. To increase the popularity of their brand these competing companies use marketing to gain access to clientele. Adidas, for example, sent every swimmer who qualified for the US Olympic trials a full bodysuit. Marketing ploys as generous as this are rare, and poor athletes from developing countries cannot afford the expensive suits.

The future

To win Olympic gold is a lifelong dream for most athletes. In Australia to win Olympic gold is to become a celebrity, a superstar, an Australian hero. The sponsorship deals that accompany the glory of gold can provide lifelong financial stability for athletes.

Olympic competitors use every advantage available to them, including the best coaches, the latest training techniques and the best equipment—this includes fast suits. There is no doubt that the companies producing fast suits will continue their research and development in order to create mass-produced fast suits for any competitor who needs to take a second or two off their personal best.

These new half suits, now called textile suits, have a place in the future of the sport. At the 2012 Olympics FINA, the international governing body of swimming, first legalised and then banned the use of full-body fast suits. Currently the rules ban full-body fast suits and allow half-body suits, allowing competitors to select from their list of approved swimsuits.

In their wisdom, FINA declared that previous Olympic and world records, all 40 of them, held by swimmers wearing fast suits will stand. If indeed the fast suits do provide the competitive advantage that FINA and all associated research has attributed to them, a return to half-body fast suits saw few swimming records broken in London 2012.

The real question for future Olympic Games is: 'How far should sophisticated technology go in order to help a professional swimmer?'.

Key concepts and definitions

Cultural impact—how culture, lifestyle, expectations and traditions may impact upon the success of an innovation.

Entrepreneurialism—the way a product is promoted.

Ethical issues—value judgements on whether something is right or wrong based on society's mores, folkways, standards and expectations.

Historical development—the evolution of innovations and emerging technologies over time.

Legal implications—what can and cannot be done to or with the innovation. These vary between countries and even states within Australia.

Useful websites

- articles.latimes.com/2009/aug/22/opinion/oe-coleman22
- science.howstuffworks.com/swimsuit-swim-faster5.htm
- sports.yahoo.com/olympics/vancouver/blog/fourth_place_medal/post/FINA-strikes-back-High-tech-suits-banned-from-s?urn=oly,178719
- www.guardian.co.uk/sport/blog/2009/mar/11/rebecca-adlington-michael-phelps-anna-kessel
- www.lattimore.id.au/2009/08/03/controversial-super-fast-swimming-suits/
- www.theage.com.au/news/sport/swimming/fast-suits-out-but-not-their-records/2009/07/25/1248457705775.html
- www.theage.com.au/news/sport/swimming/sweat-speed-and-super-suits/2009/07/25/1248457705781.html
- www.usatoday.com/sports/olympics/2009-08-02-swimming-worlds-redefining-fast_N.htm

Classroom activities

1. Using the URLs listed below and others that you may find investigate the latest research and rulings on the fast suit and its use in the Olympics and as a commercial product.
   - Swim Faster, (2003), 'The Speedo Fastskin swimsuit story', retrieved from www.swimming-faster.com/index.asp
   - Toussaint, H., 'Bodysuit yourself: but first think about it', retrieved from coachsci.sdsu.edu/swimming/bodysuit/fiveauth.htm
   - coachsci.sdsu.edu/swimming/bodysuit/Larryw1.htm
   - en.wikipedia.org/wiki/Parasitic_drag
   - www.abc.net.au/news/olympics/features/fastskin.htm
   - www.fluent.com/about/news/newsletters/04v13i1/a1.htm
   - www.sgi.com/company_info/newsroom/press_releases/2006/may/speedo.html

Sample HSC questions

Extended-response questions A pp. 215–217

Answer the following questions in relation to either the innovation and emerging technology case study you completed in class or the innovation of fast swimsuits.

1. Name and describe the innovation in this case study. List the practices used by the designers.
2. Identify and describe the factors that made the innovation a success and the factors that impacted on its development.
3. Explain how entrepreneurial activity was used to promote the innovation.
4. Describe the emerging technology that contributed to this innovation and suggest how further development may enhance the product.
5. Analyse the impact of the innovation on Australian society.
6. Draw a timeline that shows the historical development of the innovation.
7. Outline the impact of cultural influences on the innovation.
8. Evaluate the social and ethical issues associated with the innovation.
9. Evaluate the environmental impact of the innovation on society.

(15 marks each)

20 – The major design project folio: from design to solution

HSC OUTCOMES

A student:

- H4.1 identifies a need or opportunity and researches and explores ideas for design development and production of the major design project.
- H4.2 selects and uses resources responsibly and safely to realise a quality major design project.
- H4.3 evaluates the processes undertaken and the impacts of the major design project.
- H5.1 manages the development of a quality major design project.
- H5.2 selects and uses appropriate research methods and communication techniques.
- H6.1 justifies technological activities undertaken in the major design project through the study of industrial and commercial practices.

20.1 The major design project

All Design and Technology students are expected to complete a major design project (MDP). This MDP may be a product, system or environment (PSE). The MDP consists of a project and folio that together contribute sixty marks towards the Higher School Certificate examination. Each MDP is marked by a minimum of two fully-trained examiners. The project and folio are marked concurrently and given a combined score out of sixty.

Maximise your marks as follows.

- The MDP must be both of high quality and innovative in its design, construction or problem-solving. This must be present in both the project and the folio.
- Often the best projects are simple ideas that have been well developed and documented. The best advice is 'keep it simple!'.
- It is important to demonstrate problem-solving throughout the entire process and to remember that 'if you think it, write it down'. This is so the examiners can easily follow the development of your project and thinking.
- Remember that time lost can never be regained. This means that every time you sit and chat during class time, another student, at another school, is working on their project and has just moved ahead of you. Use your time wisely.
- It is necessary to demonstrate higher-order thinking skills throughout the project and folio so you must evaluate the range of solutions to problems at every opportunity.

Folio hints

Your folio needs to be divided into three sections according to the Board of Studies headings. You could do this by using a number of creative approaches. Some techniques commonly used include colour pages, dividers or title pages. The three sections should be clearly labelled as:

- Project proposal and management
- Project development and realisation
- Project evaluation.

It is important to make the folio easy for the examiners to mark. This can be done by using the Board of Studies marking criteria. Every heading must be addressed.

Major design project examination criteria

Components	Criteria	Weighting
Project proposal and management	■ Identification and exploration of the need ■ Areas of investigation ■ Criteria to evaluate success ■ Action, time and finance plans and their application	15
Project development and realisation	■ Evidence of creativity: – ideas generation, degree of difference and exploration of existing ideas ■ Consideration of design factors relevant to the MDP ■ Appropriate research and experimentation of materials, tools, techniques and testing of design solutions ■ Application of conclusions ■ Identification and justification of ideas and resources used ■ Use of communication and presentation techniques ■ Evidence and application of practical skills to produce a quality project	35
Project evaluation	■ Recording and application of evaluation procedures throughout the design project ■ Analysis and evaluation of functional and aesthetic aspects of the design ■ Final evaluation with respect to the project's impact on the individual, society and the environment ■ Relationship of the final product, system or environment to the project proposal	10

It is important to utilise cognitive organisers throughout the folio as this allows you to demonstrate how you have organised your thinking during the design process. For more information on this refer to Chapter 6. It is also necessary to provide ongoing evaluation throughout the entire folio and justify all decisions made.

When considering the quality of your MDP you will need to consider the range of skills that you will demonstrate. This is your chance to showcase the high-quality skills that you have perfected over time. If there is a small section of your practical work that you cannot complete for any reason, for example, equipment is unavailable then with your teacher's permission you may outsource that section of the project BUT it is important to note that you cannot gain any marks for work that others have completed. Instead, if possible, it is necessary to experiment to achieve quality work while managing your time to allow you to give attention to each and every detail. If a mistake or an error is made it is most important to take the time to rectify it and photograph and document it in your folio. Within the folio follow the headings below.

20.1.1 Project proposal and management

Identification and exploration of the need

Designers explore a range of needs and opportunities in order to create a way forward. This exploration may be based on their personal needs, interests, experiences and inspirations or they may encounter a problem that can be best solved through invention or innovation. The need may also develop from a second party such as a client. While some designers can work from a personal needs base it is expected that designers in companies will establish a client base. People employ designers for their skills in problem-solving and this is the role that you will take on during this course: that of a designer solving a client's problem or fulfilling their need. The need must be explored to determine that it is a real need. A designer will be given a brief from the client and then explore all avenues in the search for a solution.

Students must investigate their problem thoroughly to allow a deeper understanding of the need. This demonstrates insight to the examiner and can be achieved by identifying the associated problem and all factors contributing to or impacting on it. The information may be presented as a personal narration or as ideas in a cognitive organiser (see Chapter 6 for more detail). This depends on the type of project you choose. What is important is that you take the examiners on a 'design journey' so they can connect with your project. The best way is to 'problematise' the need using a heading of the 'design situation'. It is only when the problem is identified that a solution or design brief can be developed.

Example of a design situation for childhood obesity

With advances in technologies used for work, leisure, household and sporting activities, Australians are experiencing a more sedentary lifestyle. It is estimated that the number of obese children has doubled in Australia over the last ten years. Australian children currently experience less exercise, participate in fewer sporting activities, consume more genetically modified and fast foods, along with drinks that are high in sugar, salt and fat. The health issues associated with this are of great concern as they add to the medical costs for the Australian population.

To validate whether this need is real and that there is a problem with unhealthy lifestyles among children then statistical evidence would be required. Research is needed here to provide evidence to the examiners that the need is a real one. Examples may include statistics to back-up the stated problem of child obesity and to clarify the design situation even further.

According to the government an estimated 1.5 million people under the age of eighteen are considered overweight or obese, meaning that about twenty to twenty-five per cent of Australian children are overweight or obese. Between 1985–1997 obesity levels in the population doubled. If weight gain continues at that rate, by the year 2020 eighty per cent of all Australian adults and a third of all children will be overweight or obese. A 2009 study in Queensland showed that up to thirty per cent of Australian children have low fitness levels while sixty per cent have poor motor skills and children in general are getting less aerobic exercise. Approximately fifty per cent of obese adolescents continue to be obese as adults. It is estimated that in 1995–1996 the cost of obesity in Australia was somewhere between $680–$1239 million.

Many HSC students incorporate graphs, like the one below, to clearly communicate ideas.

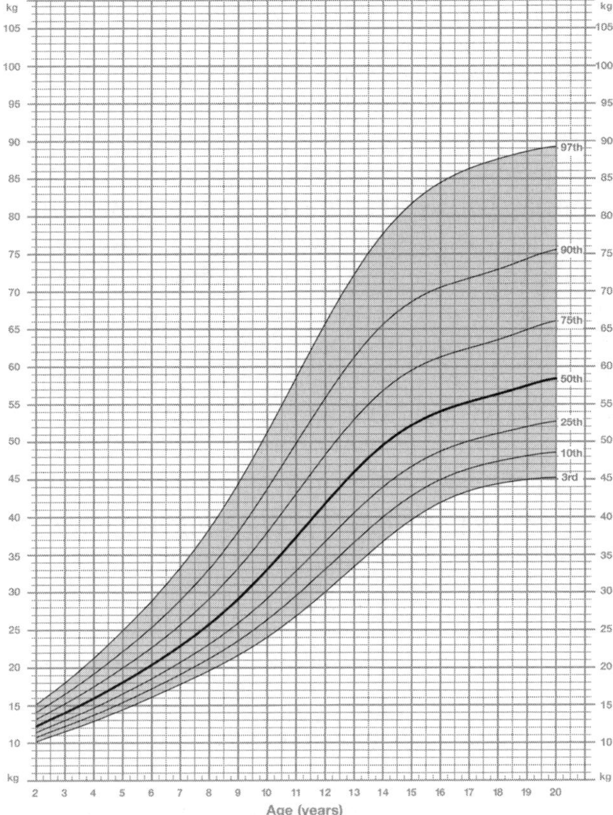

Figure 20.1 Sample graphs can communicate the need.

Once the problem is clearly researched and the design situation clarified and defined then a design brief must be written. A design brief should be kept as open as possible.

An example of a design brief based around tackling childhood obesity could be a marketing strategy that promotes a healthy lifestyle for Australian children. The marketing strategy would develop awareness of the problem in society and provide connections with health and social workers and dieticians.

Areas of investigation

Investigation is an important factor in the design process as it lists the areas that need to be pursued. Research and investigation must be carried out on materials, tools, techniques and design ideas. The areas of investigation, when completed, will further clarify the need and provide direction for further action. For example, a student investigating the problem of children being unhealthy will look at all possible causes and the sources of the data and then identify possible responses.

Research may involve surveying and interviewing children, their parents, their teachers and peers, investigating community support groups, talking to nutrition experts and looking at existing programs. Areas such as diet, exercise, lifestyle and genetics, plus available resources, facilities and costings for the project would need to be examined in the research.

The research findings provide direction for possible solutions to be further investigated on a global basis.

Criteria to evaluate success

The criteria to evaluate success links the beginning of the project to its completion, working with the proposal to create a strong image of what needs to be achieved throughout the design process. The criteria to evaluate success must be clear, practical and obtainable. A design will only become a success if the designer holds a clear idea of how to achieve that success. At this point 'success' is a prediction of what is to be expected at the end. The designer places this prediction in as a 'criteria for success' where the expectations can be checked off during the final evaluation.

The criteria for success are measured with specific parameters such as function and aesthetics. To state your criteria for success in your folio you will need to list and prioritise the functional and aesthetic properties.

Action, time and finance plans and their application

A system to maintain all the processes of the MDP will involve planning. Planning is carried out through a variety of modes. The action plan guides a series of events that lead through the design process, a time plan details how time will be managed and a finance plan involves developing a budget.

Action plan

This plan sets down a series of actions that the HSC student will need to follow from the start of the design process through to evaluation. It develops a pathway of procedures that will identify what has to happen to move toward developing a solution to the brief.

It is best to present the action plan as a cognitive organiser, for example, as a mind or concept map or a flow chart. In order to do this, select an appropriate cognitive organiser, such as the flow chart below and then add the details to each box of the steps to completion of the project.

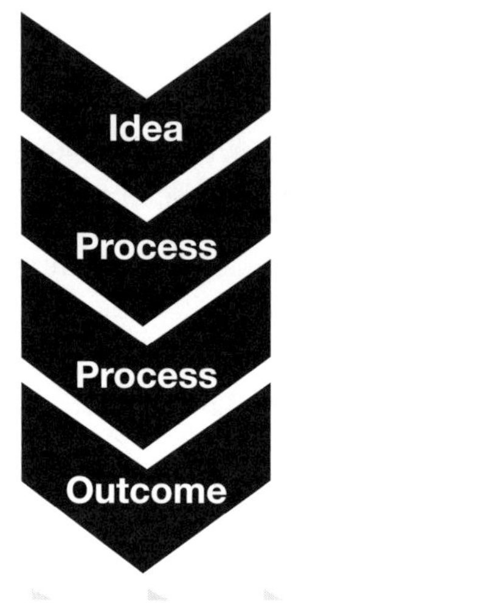

Figure 20.2 Action plan flow chart

Time plan

This provides the necessary management features to ensure everything is done on time. Specific processes might require lengthy lead times and these need to be accommodated within the plan. The time plan generally links to the action plan and provides that extra depth required when managing a project. The action/time plan can be further divided into forms of proposed and actual timing. The advantage of this is that there is a justification for any changes and this can be used in evaluations and consequent recommendations. One way to present the time plan in a cognitive organiser is to use a Ghant chart like the one below, adding an evaluation as to why the proposed and actual time plans were different.

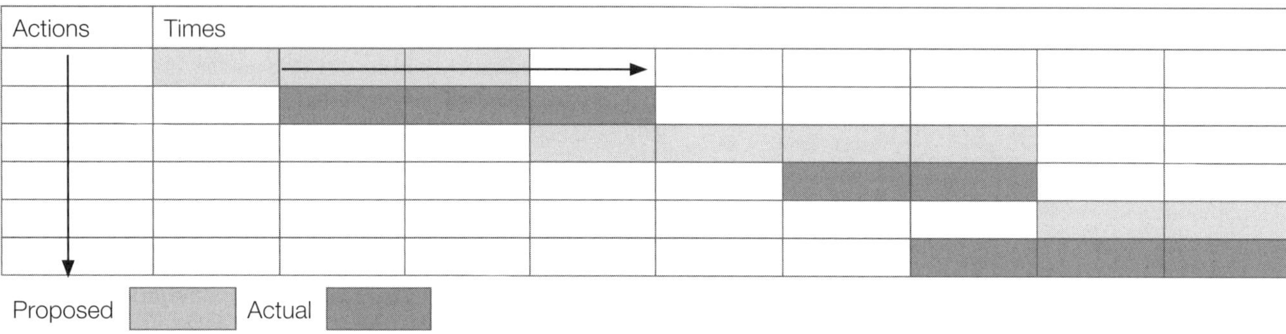

Figure 20.3 A Ghant chart

Proposed and actual times can be identified with different colours, shades or textures and you should provide a key so others can interpret your message.

Another way to present the data is to record your time plan usage each term (including an evaluation) as in the example below.

Term/week	Proposed time plan	Actual time plan	Ongoing evaluation
Term 1 Week 2	Complete the tensile strength test on …	Drew up testing experiments in the folio Term 1, Week 6. Completed the tensile test on …	Mentor was on holidays: had to await return before completing the testing
Week 3			
Week 4			

Figure 20.4 Sample time plan usage table

A third way to present the data is in a student diary. This could be done in a form similar to that below.

Term 4	1	2	3	Week 1 2 3 4 5 6 7 8 9 10
Planned to complete:				
Actually completed:				
The difference between planned and actual plan is because:				
Evaluation of the activities:				

Figure 20.5 Sample student diary

Finance plan

The management of finances is very important as it will allow a programmed spending of the budget and a 'soft landing' at the end. Financial projections can be very accurate and will avoid any cost blowouts for both the client, the designer and of course the student.

Financial plans are generally laid out in a spreadsheet where simple formulas can be used. In the example below of the finance plan for a marketing program aimed at promoting healthy lifestyles for children the student has estimated a cost of $350.

Material	Supplier	Qty	Unit	Cost	Sub	Total	Balance
				Proposed cost			$350.00
Pine for stands	Timber Express	6	L/M	$2.70	$16.20	$16.20	$333.80
Merchandising blanks	Pulse Products	23	product	$2.30	$52.90	$69.10	$280.90
Signboards	Signergy	2	each	$23.00	$46.00	$115.10	$234.90
Paint	Signergy	6	each	$8.30	$49.80	$164.90	$185.10
Ink cartridges	Ink Express	2	each	$36.00	$72.00	$236.90	$113.10
Laminating	Copyrite	15	each	$7.30	$109.50	$346.40	$3.60
Paper	Officesite	1	ream	$5.67	$5.67	$352.07	-$2.07
Phone calls	Service provider			$18.34	$18.34	$370.41	-$20.41

Figure 20.6 Sample financial plan

On this occasion, the HSC student will identify a cost blowout of $20.41. They can determine whether to allow the increase or reduce the development spending somewhere.

Another idea is to complete a budget, for the client, with income and expenditure clearly stated. It is important to explain where all finances are coming from as well as providing evidence of expenditure with receipts. The dates on the receipts may be cross-referenced by the examiner with the actual time plan.

It is the way in which these action, time and finance plans are applied that makes them a powerful and essential management tool for designers.

20.1.2 Project development and realisation

Evidence of creativity

This is a very important section of the folio. Creativity is a process involving the generation of new ideas or concepts. To give the examiners an insight into how your project developed you can creatively present a range of ideas as sketches, images, concept boards or photographs of simple models. In your folio, each existing solution needs to be recognised and then your enhancement of these ideas communicated. It is important here to show the degree of difference between the two. In this way new and existing ideas will be fully explored and clearly communicated. Original ideas can be presented as a range of different types of sketches (thumbnail, exploded, technical, production CAD) and evaluated using Edward de Bono's PMI (Plus, Minus and Interesting) which can be adjusted to Plus, Minus and Improvements.

The evolution of the design can also be shown as a series of sketches. Notations on these sketches can justify why improvements were included for each part of the design. This will allow different ideas to be explored in many ways: research, digital graphics, sketches, scanning, multimedia, and so on.

The inclusion of a high-quality final production sketch will show all different views and the labelling of functional and aesthetic features. The use of rendering, CAD or a reduced scale prototype will be an effective tool in avoiding design faults.

Consideration of design factors relevant to the MDP

Considering the design factors provides the designer with a broad brushstroke of ideas that can be factored in as part of the design development. The factors include:

- appropriateness of the design solution
- needs
- function

- aesthetics
- finance
- ergonomics
- occupational health and safety
- quality
- short- and long-term environmental consequences
- obsolescence
- life cycle analysis.

The above are examined in Chapter 11. Due to varying design ideas, HSC students would select only those factors that relate appropriately to their designing. These specific design factors would then directly determine practical and obtainable outcomes from the MDP.

The factors that will contribute to the success of an advertising campaign to promote a healthy lifestyle are listed below.

- Appropriateness of the possible design solution: the possible solution will directly relate to the target group.
- Needs: the aim of the project is to tap into the mindset of children of a specific age group in order to convince them to fight obesity.
- Safety: the end product will allow children to fight obesity in a healthy and safe way.
- Quality: the project must be of superior quality to add to the professionalism of the campaign.
- Life cycle analysis: as society is constantly changing ongoing evaluation of the campaign is a necessity: this will include analysis of the delivery time and saturation of the campaign. As with all processes there will be periods of evaluation and a reviewing of social trends.

Appropriate research, experimentation and testing

There are two types of research that can be carried out when developing your design solution. The first is used to examine solutions that are already on the market and to look for problem-solving techniques that you may incorporate into your design. This technique involves analysing information that already exists. It is second-hand information and is called secondary research. Secondary research sources include books, journals, magazines, videos, television programs and the Internet. Once this research has been gathered it must be fully referenced, read, important points highlighted and an explanation of how this research has impacted on your design thinking provided.

Another type of research is primary research. Primary research is when you carry out the research yourself. You may carry out an interview of the client or a survey to ensure that you are clear on the best solution to the need. You may practice your skills over time and produce samples to show the ongoing improvement, or create prototypes or models to ensure that designs can be created. This research is commonly used to carry out experiments on materials, using tools and experimenting with techniques. For more information refer to Chapter 7.

With any new design innovation a range of appropriate research, experimentation and testing must be undertaken. The type of testing decided upon will directly relate to the design factors associated with the design. All experiments must be relevant and have an impact on decisions made about the final design.

A scenario of an inappropriate test may involve a HSC student conducting a burn test on material being used in a project to create a new type of umbrella. Here, the more appropriate tests are about water-proofing, wind resistance, strength and durability. Burn testing would be appropriate on projects that are used in heat or fire situations such as oven mitts or fire blankets.

Ensure your testing is valid as some tests may be compromised by limited technicality or preciseness and the lack of a control. Many technical solutions can be found from existing test results or software simulations. Comparison testing is a reliable test procedure as it involves qualitative research to some degree. Test procedures can be shown in the form of text, graphics or photographs. All testing and experiments should be written up as a scientific experiment that includes an aim, method, results, conclusions and evaluation. Results may be presented as tables, charts and graphs. Include all evidence of testing, including samples, residues, remnants and techniques.

A conclusive statement explaining how each test result will impact on the decisions surrounding the design direction must be provided. It is the most important part of the experimentation as it allows you to demonstrate higher-order thinking skills. An evaluation of the tests will also provide some final details that can be developed into a course of action. Never ignore the results of your testing just because you did not like them! Testing and experimenting is considered to be a necessary and important piece of primary research when developing a project, so your conclusions based on this aspect are essential.

Application of conclusions

Once the test results have been developed and interpreted and conclusions drawn it is time to apply these conclusions to your design decisions. HSC students must identify how this primary research has benefitted the design development. At this stage it is likely that some components or areas of the design will adjust or change. This is a positive step towards a better design solution. Students should document all of the proceedings to show the design changes and must describe how each piece of research, primary and secondary, has improved the project.

Identification and justification of ideas and resources used

When any idea, resource or technology is used and incorporated into a design project it generally steers the design process. It is therefore important that all ideas and resources are identified and justified. This also provides HSC markers with 'inside knowledge' of the ideas developed and technologies investigated and used. Identification and justification may be recorded in a table where all associated information can be organised, as in the example below.

Identified resources	Justification
Teacher and mentor	The experiences help provide good advice on procedures, materials, tools and techniques
Health workers	Helping with logistics through area health services
Timber	To develop props for signage to help with advertising during Health Week
Merchandising blanks	Blanks used so designs of local appeal can be incorporated
Computer, scanner and printer	The ability to store, retrieve and process information; used to manipulate graphics to develop signage
Presentation media	Used in signage during Health Promotion Day; hand out leaflets and flyers into letterboxes; community radio advertising

Use of communication and presentation techniques

This is an area that a student can use to promote their MDP. The use of communication and presentation techniques is a powerful tool in getting across your message. Successful students always use a variety of methodologies both throughout the folio and when presenting the final MDP to the markers.

Within the folio it is important that the student engages the correct communication techniques to suit the application. For multimedia projects the use of screenshots is paramount, as might be the addition of scripting or storyboards. The addition of images depicting design development is important in projects with practical associations. Concept boards, survey results, prototypes, drawings and sketches all produce a communicating 'picture'.

When the MDP is completed and submitted for examination you are expected to display it in a way that best communicates its intention. You can provide stimuli that set the scene: photographs, drawings, samples, prototypes or, if your project has a working function, a video demonstrating its operation. You must display your MDP to its greatest advantage.

In summary, you should make the marking of the project easy for the markers by using a range of presentation and creative techniques and demonstrate the use of a range of technological skills such as word processing, spreadsheets, databases, search engines, CAD and scanned and digital images.

Evidence and application of practical skills to produce a quality project

Every MDP has a range of practical skills associated with it. A written procedure accompanied by images is needed to document step-by-step how the product was made. This will provide evidence that all work has been done with deliberation and skill by the student. A storyboard showing the progress of the design development and manufacturing processes is a great way to demonstrate your application of practical skills.

All important features, innovations and creative measures in the process of production of the project should be photographed and clearly reported. Many processes and skills may not be immediately identifiable in the MDP so providing evidence of these steps or processes is essential.

20.1.3 Evaluation

Recording and application of evaluation procedures throughout the design project

Evaluation is an ongoing process that must occur at every stage. It is important that HSC students document all forms of evaluation throughout the design process.

Ongoing evaluation provides a 'total picture' from start to finish. Recognising this in the design folio will also identify it to the marker. There are many methods of 'flagging' these evaluative stages including 'Post-it' style notes, graphic symbols or text boxes.

Analysis and evaluation of functional and aesthetic aspects of the design

Strong project evaluation only occurs when specific criteria of the functional and aesthetic aspects of the design are addressed. These functional and aesthetic aspects were identified in the section of the folio entitled 'criteria for success'.

Aesthetics—the look or psychological feel of the project needs to demonstrate a cohesive appearance.

Function—when evaluating the function of the design you must consider whether the final solution does what it was meant to do. You must describe how the project fulfills all of the functional criteria previously listed under 'criteria for success'.

The HSC student has to analyse and critically evaluate these aspects to see that what was expected actually did happen.

Final evaluation with respect to the project's impact on the individual, society and the environment

Whether it is the world of business or design studies it is critical to recognise and address any impacts a design may have on individuals, society and the environment. When evaluating the impact on the individual, an image demonstrating the product's use along with an evaluation from the client is appropriate.

Students must discuss the impact that their project has had on the individual, society and the environment. The project must be evaluated in terms of how the individual will benefit or not from its use. These may include physical, emotional and psychological benefits and problems.

A group of individuals broaden out to become a society so the MDP must also be evaluated in terms of how you dealt with societal issues, community pressure groups, minority groups and any political issues.

The impact on the environment would involve the designer considering environmental sustainability and dealing with issues such as the depletion of the planet's resources, pollution and biodegradability. Refer to Chapter 5 for more information on this topic. You may find it interesting to look at a life cycle calculator to examine how environmentally friendly your design and other designs really are. An example can be found at: www.lcacalculator.com/

The use of replacement parts and recyclable materials should also feature in your designing if appropriate. While evaluations of this sort are encouraged prior to commitment to the final design, it is also important to critically evaluate the impact on completion. It is good to include a cradle to cradle or cradle to grave analysis in this section, perhaps in the form of a flow chart.

Every MDP has varying components or weightings in this regard but the important point is the value the student places on the responsible designing and producing of their project in relation to society and the environment.

Relationship of the final product, system or environment to the project proposal

This is the moment when the designer ascertains if the final design has achieved what was expected from the start. It is here that the functional and aesthetic aspects of the design are judged against the pre-existing criteria nominated in the proposal and management sections of the design folio.

Key concepts and definitions

Design folio—a digital or A3 or A4 folder of work showing every thought and decision made during the design process, with a justification explaining why each decision was made and an evaluation explaining why it was the best decision.

Major design project (MDP)—all students complete a MDP in this course. The MDP is a project that starts with a real need or problem, follows every step of the design process and manufactures a quality solution to the problem.

Product, System, Environment (PSE)—the design solution may be presented as a product, an environment, a space which has a defined purpose or a system, process, scheme or technique.

Useful websites

- www.boardofstudies.nsw.edu.au/syllabus_hsc/pdf_doc/des-and-tech-st6-mgd-mdp.pdf
- www.cleardesignuk.com/design-brief.html
- www.hsc.csu.edu.au/design_technology/producing/projadv/2-0/2.0.3.html

Classroom activities

1. Identify and explore the need for the following designs:
   - The development of the Barangaroo site on Sydney's harbourside precinct.
   - The design of an in-flight toiletries bag.
   - An in-car navigation system.
   - Development of an e-marketplace.
2. Source three types of graphs and identify their differences and similarities and explain what each graph is displaying.

Key concept questions A p. 217

1. Why do some designers utilise graphs or statistics in their research?
2. Explain why the criteria for success is a very important tool for a designer.
3. The action/time plan is generally set up to include proposed and actual times and actions in table form. Why is this necessary for the overall planning of the project?
4. When considering design factors for your project why is it important to select only those factors that relate appropriately to your designing?
5. List the communication and presentation techniques that a designer might use in getting across their design message.
6. Why is ongoing evaluation equally important as the final evaluation?

Sample HSC questions

Objective-response questions A p. 218

Circle A, B, C or D. (1 mark each)

1. The development of a time and action plan is a wise strategy for a designer because
   A it can lead to efficient use of time and resources.
   B it can help in the achievement of set skills.
   C it can be used to work out the finances required.
   D it can be used to communicate with the client.
2. Successful completion of a MDP depends on making the link between
   A acquiring materials, using skills and a large budget.
   B investigating options, a large budget and sufficient time.
   C appropriate equipment, a large budget and using skills.
   D generating ideas, using skills and ongoing evaluation.

Short-answer question A p. 218

3. Use two design factors to explore the impact of Sydney's monorail system on the city. (2 marks)

21 —Higher School Certificate Examination techniques

The HSC Examination

The Design and Technology Higher School Certificate Exam consists of a major design project consisting of a folio and project worth sixty marks and a 1½ hour written paper worth forty marks. This chapter focuses on the written exam.

Preparation for this paper begins at the start of Year 11. Everything that is studied from then will build the knowledge and understanding needed for the exam. While the Preliminary HSC course is not examinable it forms the basis of your understandings and will be built upon in Year 12 as you develop the higher-order thinking skills of evaluation and synthesis.

As you complete your coursework and work through design projects you must ensure that you understand the specific subject language used. It is useful to start a glossary page to ensure that definitions are clear. You will also need to study your notes. Research shows that the most effective way of doing this is firstly to summarise and learn the notes and then to raise your standards of thinking using Socratic-style study groups.

Summarising notes

When learning your work you need to summarise your notes. It is best to summarise them under headings and subheadings, using dot points, and then memorise these. This process is easier if summaries are made every few days.

Each time information is summarised it should be spoken out loud, to be heard, and written down, to be seen. Seeing, hearing and speaking information encourages memory uptake. Each summary should be rewritten three times for three days, getting shorter each time. It should then be reread weekly to ensure it is not forgotten.

To ensure that your summaries have covered everything refer to the Passcards at the back of this book and also the HSC online website at www.hsc.csu.edu.au/.

Higher-order thinking skills

Exam questions are written with an expectation that students will respond by demonstrating different levels of thinking. These levels of thinking go from very simple to very complex. The higher the level of thinking that you demonstrate in your exam responses, the more marks you will achieve.

In 1954 Benjamin Bloom created Bloom's Taxonomy of Thinking. This hierarchy of Bloom's thinking is related to the marks that you receive for your HSC. The higher the level of your answer, the more marks you receive. You essay is marked according to its quality, not quantity.

Bloom's Taxonomy of Thinking

You can see that the simplest or lowest level of thinking is 'recall', so while learning your work is a great start you need to do more than this to raise the standard of your thinking, your literacy skills and ultimately your HSC mark.

In 2000 Krathwohl and Anderson revised Bloom's Taxonomy to the taxonomy illustrated below. Their research showed that it is more difficult to synthesise than to evaluate so they changed the taxonomy. Both synthesis and evaluation are considered to be higher-order thinking skills.

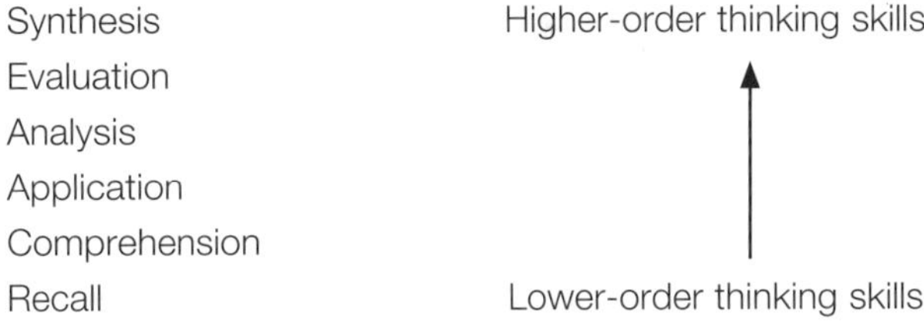

These levels can be further clarified as outlined below.

At the simplest level, or lower-order thinking, recall knowledge involves memorised responses that do not involve you adding an opinion. Lower-order thinking verbs that are used in HSC questions include list, identify, locate, memorise, name, match, review, read, reproduce, label, state a fact or recite.

Comprehension is also a lower-order thinking skill; it involves understanding, interpreting, illustrating, classifying and summarising.

Applying is a middle-level skill involving distinguishing, organising, outlining, structuring and attributing. It also applies to information involving following a written procedure.

Analysing is another middle order skill and it involves distinguishing, organising, outlining, structuring and attributing.

Evaluating is a higher-order thinking skill involving complex thought. It attracts higher marks. If you are asked to evaluate in an examination question, the easiest way to do this is to use De Bono's PMI. That is, state the positives and negatives and then synthesise suggestions for improvements.

Creating is hypothesising, designing, inventing or coming up with something new. This is problem-solving!

It is important to note here that MDPs which demonstrate synthesis (innovation in thinking or problem solving) will gain higher marks than a project that is simply beautifully made from an existing design (application). The latter would also have to be a high quality project to gain the higher marks.

What does higher-order thinking mean in the exam?

NESA provides all students with a list of verbs and a description of the thinking that is expected in responding to HSC questions. Some examples are listed in the table opposite.

Most HSC questions begin with a verb from the above list and each of these verbs relate to a level of thinking which relates directly to the marking of the question. For example, a question may ask to evaluate an object but in response some students may describe it instead. To describe is simpler than to evaluate so marks will be lower.

Throughout the Preliminary and HSC courses it is necessary to move your thinking to a higher level and for your writing to demonstrate higher-order literacy skills.

Socratic thinking

Once you know your work you need to start applying your knowledge to a variety of situations. This could be done using Socratic thinking. The great philosopher Socrates believed that answers could be found through asking questions. If he did not know the answer to a question he would ask many people for their opinion and then by taking a small part of each of their responses was able to synthesise his own ideas.

So, once you have memorised your summaries you should start to ask your peers and teachers questions about issues related to the topic. Many students form Socratic study groups to help raise their thinking skills, and also their marks.

How to form a Socratic study group

- Form a study group with three or four other students with similar abilities and dedication to you.
- Divide topics up between the group members.
- Each group member is to research and become an expert or group leader on their topic. They should write four to five questions (teacher's help permissible) that take their thinking beyond what is expected in the classroom.
- The group should meet regularly to study each topic, with the leader guiding the discussion with previously planned questions and possible responses.
- Participating in a study group will help develop your higher-order thinking skills.

Written examination hints

- Be well organised and prepared. Good preparation begins in Year 11 and continues all through Year 12.
- Study in the morning rather than the afternoon and evening when you are more likely to be tired.
- Alternate intellectual and physical activity. Study a topic for fifty minutes maximum and then take a ten-minute break and do some physical exercise to allow your brain time to absorb what you have considered.
- List the things that are hard to remember on pieces of paper or notes and paste them on a wall where you will read them frequently, for example, on the toilet door.
- Read these notes just before you walk into the exam.
- Reduce exam stress by avoiding people who are overanxious or happy before the exam; stay calm.
- Take a bottle of water into the exam with you.
- Write an essay plan before writing an extended response.
- Know your work!

Term	Definition
Account	Account for: state reasons for, report on Give an account of: narrate a series of events or transactions
Analyse	Identify components and the relationship between them; draw out and relate implications
Apply	Use, utilise, employ in a particular situation
Appreciate	Make a judgement about the value of
Assess	Make a judgement of value, quality, outcomes, results or size
Calculate	Ascertain/determine from given facts, figures or information
Clarify	Make clear or plain
Classify	Arrange or include in classes/categories
Compare	Show how things are similar or different
Construct	Make; build; put together items or arguments
Contrast	Show how things are different or opposite
Critically (analyse/ evaluate)	Add a degree or level of accuracy, depth, knowledge and understanding, logic, questioning, reflection and quality to (analyse/evaluate)
Deduce	Draw conclusions
Define	State meaning and identify essential qualities
Demonstrate	Show by example
Describe	Provide characteristics and features
Discuss	Identify issues and provide points for and/or against
Distinguish	Recognise or note/indicate as being distinct or different from; to note differences between
Evaluate	Make a judgement based on criteria; determine the value of
Examine	Inquire into
Explain	Relate cause and effect; make the relationships between things evident; provide why and/or how
Extract	Choose relevant and/or appropriate details
Extrapolate	Infer from what is known
Identify	Recognise and name
Interpret	Draw meaning from
Investigate	Plan, inquire into and draw conclusions about
Justify	Support an argument or conclusion
Outline	Sketch in general terms; indicate the main features of
Predict	Suggest what may happen based on available information
Propose	Put forward (for example, a point of view, idea, argument, suggestion) for consideration or action
Recall	Present remembered ideas, facts or experiences
Recommend	Provide reasons in favour
Recount	Retell a series of events
Summarise	Express concisely the relevant details
Synthesise	Put together various elements to make a whole

The Design and Technology HSC Examination specifications

The HSC Examination consists of a written paper worth 40 marks and a major design project worth 60 marks.

Time allowed: 1 hour 30 minutes plus five minutes reading time.

The paper will consist of three sections.

Section I (10 marks)
There will be ten objective-response questions to the value of ten marks.

Each question will have four options: A, B, C and D. Of these, one will be more correct than any of the others. When answering you should read the question and cover the answers. Then you should consider what the answer is before matching it to the correct option. Allow approximately 15 minutes for this section of the paper.

Section II (15 marks)
There will be a number of short-answer questions to the value of 15 marks. Questions may contain parts and there will be approximately four items in total.

At least one item will be worth from four to six marks.

In this section it is possible that you will be given a piece of stimulus material with the questions relating to this. The questions will be based on you knowing what the terms mean and applying them to the material. For example, 'read the stimulus provided and explain what the term "intellectual property" is referring to'. This means that you will need to know what the term means and explain its application in this situation.

Allow approximately 30 minutes to answer this section.

Section III (15 marks)
There will be one extended-response question. The question will have an expected length of response of around four pages of an examination writing booklet (approximately 600 words).

The extended-response question is worth 15 marks and there is one mandatory question. Allow approximately 45 minutes to answer this question. Your answer must be well planned; the plan may be written into the response booklet before you attempt to answer the question.

Here is an example of an extended-response question:

For an Australian innovation that you have studied, discuss all aspects of at least two ethical issues surrounding the design.

When planning the response follow the steps below:

1. Circle the verb. This is done to ensure that you answer in a way that will maximise your marks. To discuss, as defined by NESA, is to identify issues and discuss points both for and against.
2. Underline the topics to be discussed.
3. Plan the essay.
4. In the first paragraph, define all terms circled or underlined, for example:
   - discussion will identify issues and discuss points both for and against.
   - Australian innovation will need a brief but detailed description of the innovation with an explanation of the emerging technology used and an explanation as to why it is innovative.
   - ethical issues: at this point two ethical issues will be introduced.
5. Paragraph two is a description of one ethical issue.
6. Paragraph three discusses the positives of this ethical dilemma.
7. Paragraph four discusses the negatives of this dilemma.
8. Paragraph five is a description of a second ethical issue.
9. Paragraph six discusses the positives of this ethical dilemma.
10. Paragraph seven discusses the negatives of this dilemma.
11. Paragraph eight is the conclusion which includes a statement that ties it all together.

Key concepts and definitions

Higher-order thinking skills—the complex level of thinking that allows students to gain the most marks. This may involve synthesis and/or evaluation.

Metalanguage—thinking about the language and terms used in a specific situation, for example, 'explain' is a verb commonly used in HSC questions.

Socratic thinking—a type of thinking that is based on questioning to promote higher-order thinking.

Taxonomy or hierarchy—levels of difficulty.

22 – Sample HSC Examinations

Sample HSC Examination 1

Section I: Objective-response questions

Total marks 10
Attempt Questions 1–10.
Allow about 15 minutes for this section.

1. Working collaboratively in a design field would involve
   A a team member presenting a prototype.
   B all team members sharing ideas over a design sketch.
   C the team leader presenting a PowerPoint presentation.
   D each member of the design group taking a turn at presenting a model.

2. A product developed for global acceptance is best described as
   A a product designed by a design team from overseas.
   B a product that is produced off shore.
   C a product that relies on minimal design variation.
   D a product using technologies from varying international sources.

3. The design of a pedal-powered grain mill to be used by impoverished farmers in developing countries is an example of a design solving local challenges. Using this low cost, low-tech technology is an example of
   A appropriate design.
   B recycled design.
   C cost-effective design.
   D renewable design.

4. The projection of a product into the marketplace was assessed over a period of four years. The graph below shows the results of this assessment.

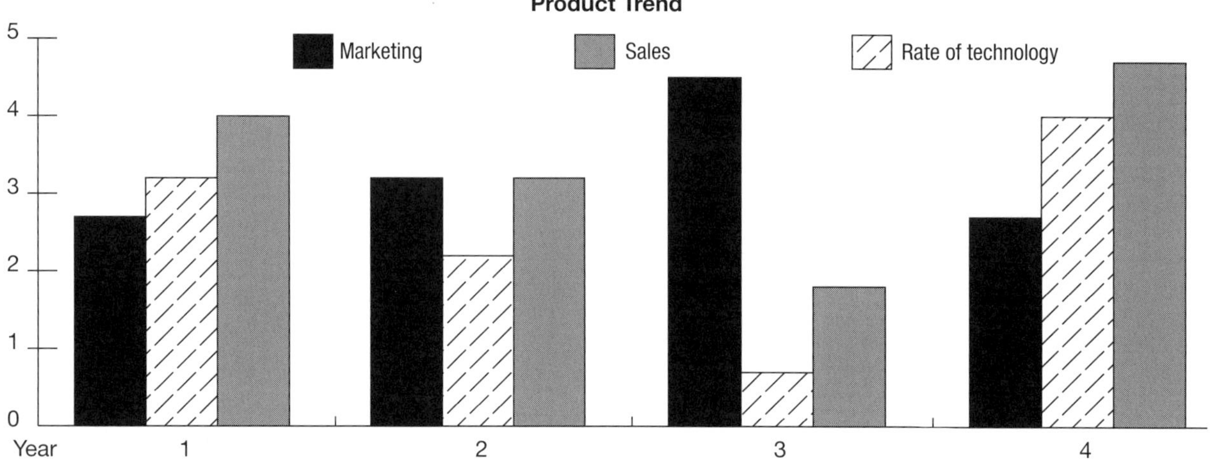

   The assessment proved that
   A the product grew in popularity over time.
   B new technology resulted in increasing sales.
   C a marketing strategy was essential to the product's success.
   D the product was placed into a new market environment.

5. A designer's idea is best kept safe by
   A selling it through a verified broker.
   B establishing intellectual property.
   C storing it with a security firm.
   D early release of the design to be in front of the competition.
6. In what way would 'green design' be used to contribute most effectively to the environment?
   A use of a sustainable power source
   B use of an energy star rating in the designing
   C use of dissimilar materials in the design
   D use of energy efficiency in the manufacturing process
7. A designer has been commissioned to design a modern swimming complex. What would be the most ethical design factor that they would need to consider?
   A keeping the entry prices low
   B guard rails and non-slip surfaces
   C secure changeroom facilities
   D recycled water for the pools
8. Which of the following lists of functional criteria is the most appropriate when developing a new backpack for bushwalking?
   A appearance, size, and cost
   B cost, appearance and ease of use
   C weight, size and ease of use
   D colour, weight and size
9. Which of the following terms best describes the development of an innovation into the marketplace?
   A commercialisation
   B marketing
   C life cycle assessment
   A product introduction
10. A large company is planning on designing and producing Australian-made souvenirs for the Japanese tourist market. What initial action would the company take in developing this product?
   A Research and identify any cultural sensitivity.
   B Research the materials and the machine tooling.
   C Research and identify the culture and lifestyle of Japan.
   D Research the cost of similar articles in souvenir shops.

End of Section I

Section II: Short-answer responses

15 marks
Attempt ALL parts of Question 11.
Allow about 35 minutes for this section.
Answer the questions in the spaces provided.

Question 11 (15 marks)

The latest digital radio, packed full of design features, is set to change the face of radio. Via this Internet-connected device, users are able to tap into social networking sites such as Facebook and twitter. This update of the humble clock radio will change the radio landscape by harnessing the social networking phenomena and delivering it on an interactive radio.

It has been claimed that the Sensia will do for radio what the iPhone did for the mobile phone. Via the dedicated Internet portal, users will be able to buy and download applications. These may eventually include anything from tools allowing the monitoring of electricity use to home security functions.

(a) Outline TWO reasons why the Sensia digital radio can be considered an appropriate technology for contemporary Australian society. (2 marks)

(i) ______________________________

____

____

(ii) ______________________________

____

____

(b) Identify why the new Sensia digital radio is considered an innovation. (2 marks)

____

____

____

Question 11 (continued)

(c) Outline the process of the following design factors as they relate to the Sensia digital radio. (3 marks)

Cost	
Function	
Aesthetics	

(d) Define the importance of the following factors in the success of the Sensia digital radio. (2 marks)

(i) Timing

(ii) Available and emerging technologies

(e) Discuss why the Australian Government and its policies are becoming more active in the control of this style of technology. (2 marks)

(f) Explain how commercial and industrial agencies would have encouraged entrepreneurial activity that helped develop the Sensia digital radio. (4 marks)

End of Section II

Section III: Extended-response question

Total marks 15
Attempt Question 12.
Allow about 40 minutes for this section.

Question 12 (15 marks)

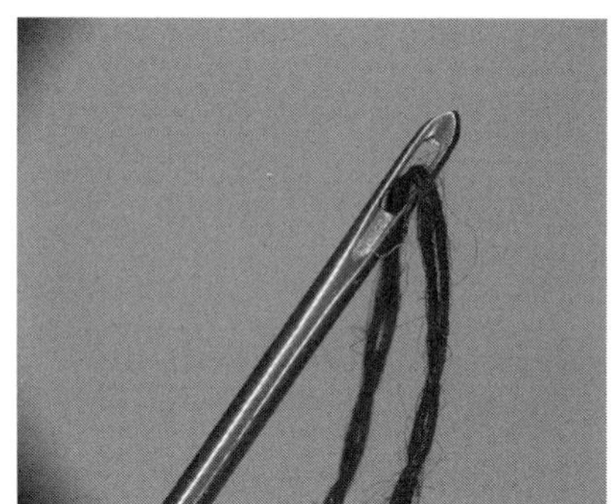

In response to the need to photograph close-up images clearly, an amateur photographer designed a camera lens for extreme close-up shots. With the use of a lens adapter and a ring light, the photographer was able to capture images in minute detail.

Analyse the innovation of the camera and identify its development towards commercial success.

Sample HSC Examination 2

Section I: Objective-response questions

Total marks 10
Attempt Questions 1–10.
Allow about 15 minutes for this section.

1. Design is best described as
   A a decorative process.
   B the functional and aesthetic attributes of a product.
   C a way of improving an existing product, system or environment.
   D a lifelong, experiential learning process involving problem-solving.
2. The design process involves a number of steps including
   A create, make, evaluate.
   B divergent and convergent thinking.
   C sketch initial ideas, research, test, experiment, final sketch, manufacture and evaluate.
   D sketch initial ideas, research, test, experiment, final sketch, manufacture with ongoing evaluation.
3. Responsible design refers to
   A a designer's corporate responsibility.
   B a designer's personal responsibility.
   C a designer's response to the impact of their designs on the natural world.
   D designers considering the impact of resource choice, manufacturing techniques, transportation and marketing on the natural world.
4. The factors affecting designing and producing include
   A the factors that will inhibit the product's success.
   B the factors that will contribute to the product's success.
   C only the factors that are appropriate and relevant to the design situation.
   D all of the following: appropriateness of the design solution, needs, function, aesthetics, finance, ergonomics, occupational health and safety, quality, short- and long-term environmental consequences, obsolescence and life cycle analysis.
5. Which of the following should an ethical designer consider as being a key and integral part of the design process?
   A a life cycle analysis
   B the recycling of materials
   C the raw materials being used
   D the energy required to manufacture the product
6. Which of the following should be carried out by a responsible designer when designing a new product?
   A Creatively resolve issues that existing products do not address.
   B Consider their corporate and personal responsibility.
   C Respond to the impact of their designs on the natural world.
   D Consider the impact of resource choice, manufacturing techniques, transportation and marketing on the natural world.

7. A designer has been asked to design a system to re-use household waste. When developing a design concept, which of the following will the designer need to consider?
   A aesthetic qualities
   B local government regulations
   C how recycled materials can be utilised
   D the impact of pollution on the air, water and land

8. Innovations and technological change assist in the development of new products. Which design consideration is the most significant?
   A The design can be manufactured.
   B The design is safe and can be easily used.
   C The product uses only new technology.
   D The users understand how the new technology works.

9. Sustainability is a global issue that aims to ensure that
   A the hole in the ozone layer is reduced.
   B natural resources can be utilised by future generations.
   C all materials used in production are recyclable, reusable or renewable.
   D new designs do not contribute to global warming, and pollution is minimised.

10. Why is ongoing project evaluation important when designing and creating?
   A It saves the designer time, energy and money.
   B It allows the designer to get feedback at each stage.
   C It ensures the designer adheres to their time and action plans.
   D It ensures the final project will meet the needs of the design brief.

End of Section I

Section II: Short-answer responses

15 marks
Attempt ALL parts of Question 11.
Allow about 35 minutes for this section.
Answer the questions in the spaces provided.

Question 11 (15 marks)

During the development of design solutions, designers and clients must communicate clearly as solutions are visualised and design decisions made.

(a) List FOUR ways in which design solutions are visualised. (2 marks)

(i) ______________________________

(ii) ______________________________

(iii) ______________________________

(iv) ______________________________

(b) Explain why in the designer/client relationship clear communication is of the utmost importance. (3 marks)

The designer analyses the problem, considers the design brief with all limitations, researches the latest design ideas, materials, tools and techniques and then visualises a solution which is shared with the client. The intellectual property of this design belongs to the designer.

(c) What is meant by the term 'intellectual property'? (2 marks)

An ethical dilemma may occur if the client believes that they have contributed to over fifty per cent of the design ideas.

(d) Who would own the intellectual property in this instance? Justify your answer. (3 marks)

(e) Explain one way that designers protect their intellectual property. (2 marks)

(f) Once a design solution is visualised, prototypes and models are developed. Propose and explain one research method that could be used to further improve this evolving design solution. (3 marks)

End of Section II

Section III: Extended-response question

Total marks 15
Attempt Question 12.
Allow about 40 minutes for this section.

Question 12 (15 marks)

The impact of environmental change on the planet cannot be denied. While the cause and possible future effects are debatable, designers must act in a responsible and ethical manner. This will ensure the sustainability of the planet for future generations.

Describe the causes of environmental change and their impact on the planet, and predict ways in which responsible designers may overcome these global issues by promoting sustainable design solutions.

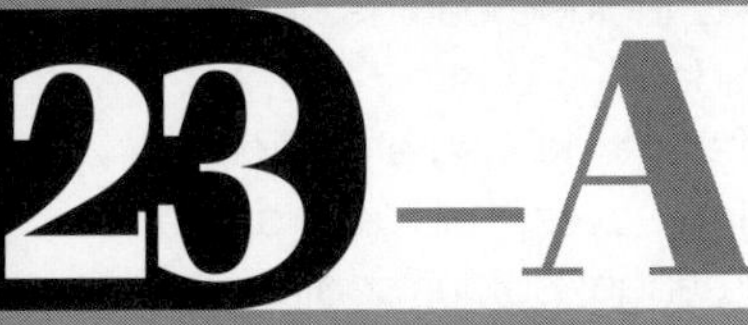

nswers to end-of-chapter questions

Chapter 1 p. 16

Key concept question

1. In your comparison include information on the following:
   - name of designer
   - designer's educational background/business training
   - factors that contribute to the designer's success
   - designer's special signature (design expertise)
   - career highlights and low points
   - employment history
   - name and description of current company/ies
   - include images and a description of products produced at the company
   - design process used
   - entrepreneurialism
   - designer environmental policies and philosophy.

Sample Preliminary questions

1. In order to maximise your marks for this question you need to define the terms 'design', 'problem-solving' and 'experiential learning'. You need to discuss how design is not about aesthetics or decoration but is a process used to solve problems creatively and also discuss the design process and how finding solutions involves using new and emerging materials and technologies. You must then justify how the preceding points require the use of design. (15 marks)
2. You need to define 'design', 'technology' and the 'design process'. You must explain the importance of integrating higher-order thinking, creativity, innovation and design as a conceptual tool used to solve problems. Points that should be covered are listed below.
   - Designing is a complex intellectual activity that requires higher-order thinking skills; it is a conceptual change.
   - Design, in this instance, refers to a creative synthesis of ideas using a design process and technology being the repetitive use of tools to complete the manufacture of an item.
   - Design processes are used to create synthesised, original and purposeful solutions to problems.
   - Design is used to 'make meaning'.
   - Design promotes lifelong learning.
   - Technology education is an area requiring low-level intellectual skills and is used as a vehicle for developing physical skills in manipulating materials.
   - Designers must use technology to realise solutions to problems. (15 marks)

Chapter 2 pp. 24–26

Key concept questions

1. A safety poster range may include back safety; resuscitation; first-aid guide blue poster; first-aid guide red poster; fire safety; forklift safety; slips, trips and falls; ladder safety; eyes, hands and hearing; tools; bites and stings; colour coding or sun safety. All safety posters should include:
   - a 'catchy' title in a large, coloured font
   - a simple, clear and coloured graphic demonstrating correct procedures
   - three main short points in a large font that clearly explain the main thrust of the poster
   - up to three additional points in a smaller font that support the thrust of the poster
   - a link to a website or address where additional information can be obtained
   - a range of headings, colours and fonts to emphasise points.
2. A sample answer, for manufacturing a textile item, is set out below.

Procedural steps	Addressing safety issues
1. Cut out paper pattern	1. Lay pattern on table to ensure it is flat and only the pattern can be cut.
2. Pin pattern to fabric	2. Place heads of pins on edge of fabric, all lying in same direction to ensure no stabbing by pins. Use a thimble to reduce damage to fingers.
3. Cut out pattern	3. Lay flat on table to ensure straight edges are cut and nothing else gets caught in the scissors. Ensure the scissors are sharp so longer cuts, resulting in less damage, can occur.
4. Pin pattern pieces together	4. Place heads of pins on edge of fabric, all lying in same direction to reduce damage to machine and eyes if a pin is stitched over.
5. Tack using hand stitching	5. Use a thimble to reduce damage to fingers.
6. Machine pieces together	6. Pre-test stitch dimensions and tension on a scrap to ensure threading is correct and machine is stitching correctly.

3.

Product/system or environment	Aesthetic properties	Functional properties
Sydney Harbour Bridge	The steel frame gives the appearance and feel of strength, rigidity and safety The curved wave of steel over the bridge softens the appearance It blends in with the harbour surrounds	Strong, sturdy and safe It carries both automobile and rail traffic in both directions Changing and movable lanes allow the bridge to cope with peak hour traffic
Sydney Opera House	Aesthetics fit in with the harbour environment The Opera House resembles sails on the water Nautical colour scheme of white sails and blue harbour and an earthy coloured base	Enables quality sound performances Allows a number of performances to occur concurrently at a number of venues within the structure Includes disabled access Includes storage space for costumes, sets and scenery
Backyard pool surrounds	Provides a relaxing atmosphere Provides a harmonious colour scheme Creates an atmosphere using palm trees and furniture synonymous with leisure	Furniture allows people to rest Pool does not have surrounds that can be tripped over Provides a dry area for children to play though having no pool fence makes it undesirable for young children
Formal gown	Elements and principles of design work together to create a style of elegance Harmonious colour scheme creates a classic look in this outfit Fabric texture creates the desired effect	Dress style features make it a 'show-stopping' attention-getter The fitted bodice and flowing skirt create an outfit that shows off the body on the red carpet The fitted bodice emphasises a shapely bust and narrow waist
Snowboard	Able to be purchased in a variety of colours and motifs Generally recognised as works of art	Strong connectivity between the board and the rider Shapely design to help with speed and manoeuvrability
'Aqua' aftershave bottle	The aqua colour of water promotes a gentle but masculine feel The see-through package shows aqua-coloured scent that resembles clear ocean water The clear square glass container has a strong and solid feel which gives the impression of classiness The square lid is a strong masculine shape that contrasts with the softness of the aqua liquid The name, font and colour of the logo could be used to create the 'image' The silver lid promotes the high quality of the product	The aftershave is a subtle scent that is understated and masculine just like the package The package contains the perfume, it is see-through as the scent can withstand light, and is sealed from air to stop the scent deteriorating
Poached pears	Food is presented using a colour scheme that makes it appeal to the client's appetite Texture is used to make the food more appealing	Textures and flavours blend to make the food appetising

4. Mobile phones are planned for obsolescence in the following ways.

- Technical obsolescence: the battery in a mobile phone lasts only about 500 cycles before losing large amounts of its capacity. Production of these batteries is usually stopped at around the same time as the product is discontinued, rendering the product worthless once the batteries wear out.
- Style obsolescence: the style of the mobile phone is designed to last one to two years before going out of fashion.
- Functional obsolescence: a mobile phone is not just a telephone—most now include a range of ever-expanding functions such as a radio, telephone directory, message service, Internet access and satellite navigation.
- Value engineering: the mobile phone is ever-increasing in its functions. Each new function involves research and additional costing. In order to keep the mobile phone at an acceptable cost these innovations are revealed annually.

Sample Preliminary questions

Extended-response questions

1. The factors influencing the design's success are outlined below.
   - Appropriateness of the design solution: the chair reflects the millennium. Made from modern materials and uses an innovative technique. The lounge was not created for comfort but to be displayed in an art gallery to reflect the year 2000. It is an appropriate design solution to the design brief.
   - Needs: the need was to create an artistic piece of furniture with a millennium theme that would be both functional and aesthetically pleasing.
   - Function: the piece of furniture is functional in that people can sit on it while examining other art pieces or while waiting, but it is not comfortable over extended periods of time. The function suits the purpose that the chair was created for.
   - Aesthetics: the aesthetics promote a millennium feel through the use of materials that have a shiny, metallic, studded appearance. The design has a 'knight in shining armour' appeal but has been recreated to have a modern, futuristic look.
   - Elements of design: curved lines and shape create a comfortable feel. The shape is that of a reclining human body which contributes to the relaxation effect. The direction of the line is horizontal so that it creates a relaxing or sleeping image. The shiny colour and metallic texture contribute to the modern millennium effect.
   - Finance: the cost of this chair is very high as it was a one-off original creation, not a mass-produced product.
   - Ergonomics: the ergonomics of the reclining body was extremely important as the chair is shaped to resemble this posture. It is assumed that a body would slide into the design and be completely supported by the lounge. It is only on closer inspection that the studs can be seen and the comfort reconsidered.
   - Occupational health and safety: when manufacturing this product it was essential to ensure all workplace safety laws and Acts were followed to ensure a minimal hazard risk.
   - Quality: the design specifications were clearly set out in a production sketch, and then manufactured by hand by the designer to ensure the product was of superior quality.
   - Obsolescence: this design did not use planned obsolescence as it was a one-off piece. However, the miniature replicas are considered works of art for people to invest in.
   - Life cycle analysis: this product was not meant for mass production so the environmental impact and ecological footprint is not as great as it could have been. Environmental costs involved the extraction and processing of the stainless steel, transportation to the manufacturing site, the use of fossil fuels during the manufacturing process, the waste produced during extraction, transportation and manufacturing, and the distribution of the product to the required site. (15 marks)

2. Obsolescence is the state of being which occurs when a person, object or service is no longer wanted even though it may still be in good working order. Obsolescence frequently occurs because a replacement has become available that is superior in one or more aspects. Obsolescence also occurs as new functions and technology are introduced to products. These are often introduced over a number of years rather than all at once to allow consumers access to new technology one step at a time at reasonable prices. The desire for new technological functions drives the consumer to constantly upgrade their product to the newest on the market. In order to answer this question you need to provide information on each of the following types of obsolescence:
   - technical obsolescence
   - functional obsolescence
   - planned obsolescence
   - planned systemic obsolescence
   - style obsolescence
   - notification obsolescence
   - obsolescence and durability
   - postponement obsolescence. (15 marks)

Chapter 3 pp. 37–38

Key concept questions

1. Architecture: the effect is one of strength and stability created by vertical lines at ninety-degree angles. The vertical erect lines seem to extend upwards beyond human reach towards the sky. These lines often dominate in public architecture and corporate headquarters. Extended perpendicular lines suggest an overpowering grandeur beyond ordinary human measure.

   Furniture: the effect created is one of solidity. The horizontal lines are in a curved shape and are used to soften the straight vertical lines. They combine to communicate stability and solidity. Rectilinear forms are stable in relation to gravity and are not likely to tip over. This stability suggests permanence, reliability and safety.

   Fashion: the gown has a feeling of elegance created by diagonal lines. Diagonal lines suggest a feeling of movement and direction. The curve of these lines around the body creates a sensual effect. This reinforces the sense of movement as the eye fluidly follows to the focal point of the narrowest point on the legs from which the fishtail graduates out.

2. Harmonious colour combinations are soothing and easy on the eye. This scheme produces a rich, peaceful and pleasing effect. Contrasting colour combinations attract attention and stand out.

3.

Element or principle	Fashion design	Dubai architecture
Line and direction	Vertical, gently curving lines are reinforced around the hips and neckline and throughout the design	The diagonal line provides the impression of movement
Shape and size	The shape is very feminine, following the curves of the female body; this gives a sensual effect	The shape is that of a sailing boat that dominates the landscape
Colour	A single white hue is used; this has a strong intensity that promotes a feeling of purity	The monochromatic blue and white colour scheme, using shades and tints of blues, reinforces the 'sailing on the water' imagery
Texture	A smooth shiny texture is used; this shows movement in the design as the light reflects across it	A light smooth texture is created through the use of glass and strong structural beams
Proportion	This design uses the golden mean ratio; the fishtail feature is proportionate to the entire gown	The building is divided into five sections and uses the golden mean to proportion it
Balance	This dress is formally or symmetrically balanced because if a line of symmetry is drawn down the middle it is identical on both sides	This building is informally or asymmetrically balanced which adds to the sailing and movement theme
Repetition	Repetition has been used in the curved lines in the design	Horizontal lines are used repetitively to ensure a stable, restful and safe feeling at this hotel, office and apartment block
Contrast	The contrast in this design occurs where the fishtail joins the dress; here curved lines give way to straight diagonal lines that create movement	The contrast occurs where there are multiple minor horizontal lines counterbalanced with a strong curved line which makes the design interesting
Emphasis	The focal point of this design is the neckline, waist and hips, emphasising a womanly hourglass figure	The focal point of this design is the sail
Rhythm	The shine of the dress as the wearer moves creates the rhythm of this design	The rhythm is smooth as the eye slides down the sail before looking for and at the details
Unity	The unified effect is one of softness and sensuality	The unified effect of this design is one where the building successfully looks like a sailboat that is sitting on a constructed island

Unity is the culmination or final point of a design. Unity is achieved within a design when every component, detail, feature and element and principle of design supports the central concept and overall quality, creating a sense of completeness.

Question 4 is based on your individual design work. However, the language used to link elements and principles and the psychological aspects of the design are referred to throughout this chapter so check to ensure that you have applied the correct terms and descriptions.

Sample Preliminary questions

Short-answer questions

1. Good design creates a positive physical, emotional and psychological comfort and addresses functional and aesthetic considerations. Great designers consider each and every element and principle of design before finalising their ideas as these are the building blocks of great design. Each element reinforces the psychological aspect of the design. Poor designs demonstrate conflicting elements or unclear principles such as a combination of straight and curved lines. (2 marks)
2. Functional properties are aspects that are used to make the design suit a specific purpose. Terms used to describe function include 'strong', 'comfortable', 'relaxing', 'sturdy', 'tactile', 'hard wearing', 'ergonomically friendly' and 'durable'. Aesthetic properties are decorative details that could be removed without affecting the item's function. Terms used to describe aesthetics include 'pretty', 'tactile', 'smooth', 'bubbly', 'intricate', 'elaborate', 'complicated', 'simple' and 'convoluted'. (2 marks)
3. The elements of design include line and direction, shape and size, value, texture and colour. The principles of design include proportion, balance, rhythm, emphasis or dominance, contrast, harmony, repetition and unity. (2 marks)

Chapter 4 p. 50

Key concept questions

1. In this question you have to 'outline', which means you must sketch in general terms and indicate the main features of the popularity of the cottage industry. The cottage industry was very common in the time prior to the Industrial Revolution when a large proportion of the population was engaged in agricultural activities as it provided a way for farmers and their families to earn extra money during the slow winter months. Many of those who worked in the cottages became quite skilful and these skills were the beginning of specialisation and the introduction of artisans.
2. To 'examine' the idea of the 'modern-day cottage industry experience' you need to inquire into the use of this structure in today's context. In modern terms 'cottage industry' describes a variety of home-based employment activities. People see this as an extended hobby grown out of a love of manufacturing and an opportunity to follow an alternative way of life. Even today there are many local productions of traditional handicrafts being developed. Telecommuting is a technological form of the cottage industry idea.
3. To 'outline' this home-based activity you must sketch it in general terms and indicate any main features of the work. Manufacturing companies sub-contract their garment making out to textile workers in their homes. Outworker welfare is strictly controlled by the government. All major garment manufacturers currently employing outworkers ensure that their workers are dealt with with a high degree of ethics, something which is also becoming part of their marketing campaigns.
4. You need to identify and name these Acts. They are the *Occupational Health and Safety Act 1989*, which is the *Prevention Act*, and the *Accident Compensation Act 1985* which compensates for injury.
5. To 'describe' the three manufacturing techniques you are required to provide characteristics and features for each of them. This will help you understand the techniques and distinguish between them.

   Custom production is a one-off manufacturing technique used in a similar manner to hand-crafted or custom developed processes. A special order is placed and several people may work on the product. Customising allows a focus on the exactness and uniqueness of the detail.

   Batch or job-lot production is used when a production run has limited numbers. Machines are set up to suit the conditions and numbers for the batch. This cost-saving method is very popular in small industries.

   Continuous production is used when demand is high. It allows for a steady supply of a product, often using an assembly line production method.
6. To 'describe' the three management techniques you need to write about the characteristics and features of each technique.

   Total quality management or TQM is a management strategy aimed at ensuring all people in an organisation understand the importance of quality.

Value-based management or VBM ensures corporations are managed consistently on value in terms of investment, cost and quality. A strong corporate culture is shared amongst the employees.

Just-in-time management or JIT is an inventory system designed to ensure that materials or supplies arrive when required to reduce storage and holding costs. The system relies on customer demand and is used in most mass-production assembly plants where the supply of goods is crucial.

Sample Preliminary question

Extended-response question

1. (a) In this question you have to 'explain', which means you must provide a 'why and how' to the design and production of your MDP. Identify the production activities and link these to the resources available and skill level required. Identify the importance of customisation and the need to use some discrete technologies. Identify the management methods used to bring your project together in terms of both time and cost. (10 marks)

   (b) In this question you have to 'predict', which means you must suggest what may happen if your MDP was industrially produced. This assessment should be based on current available information. Try to visualise your product to the point where you can see it being used everywhere. Trace back from here to the point of manufacture and the industry that supports this. Your design will depart from its original individual status and be transformed by mass production techniques. Factory management techniques, business expansion opportunities and shareholders may all be part of the business plan. (5 marks)

Chapter 5 pp. 60–61

Key concept questions

1. Examples could include happiness, honesty, integrity, wealth, freedom, status, security, laughter, quality, friendship, hygiene, employment, leisure and time.
2. An example could be that in Indigenous Australian culture to look someone in the eye when you speak to them shows disrespect.
3. Ensure that all sketches are produced using quality sketching techniques that are fully labelled for both function and aesthetics and then justify how they solve the problem.

Sample Preliminary questions

Objective-response questions

1. **D** is the correct option.
2. **A** is the correct option. (1 mark each)

Short-answer questions

3. Cradle to grave has a finite life and usually ends in waste disposal. Include an example.

   Cradle to cradle means a product can be reused and recycled in an ongoing manner. Include an example. (2 marks)
4. Forests are the lungs of the planet, providing oxygen, and they are rapidly being cut down in order to grow crops to feed the earth's ever-increasing population. (3 marks)
5. The debate surrounds the cause and the cure. Some believe it is a natural phenomenon but most scientific investigation supports the theory that it is caused by the use of technology by humans. (4 marks)

Extended-response questions

6. For your chosen design project, include the following criteria in your answer:

(a) Describe your design solution in detail and thoroughly investigate the problem. (3 marks)

(b) Include applicable data on personal, social and community values and explain how they relate to the client and the design solution. (2 marks))

(c) An individual problem is one that is concerned with someone's personal needs and wants while a community-based problem reflects the needs of a group of people within a community, for example, the local cricket club or a charity organisation. (4 marks)

(d) Equity is concerned with fairness to all: the client, the designer, the resource supplier, the transportation supplier, the manufacturer, the producer and the consumer. (6 marks)

7. (a) Describe the problem and the solution with a fully-labelled production sketch and also in terms of the needs, solutions, functional and aesthetic properties, technologies used in production and innovation in the design solution. (1 mark)
   (b) Explain how gases may have been produced that contributed to global warming during the resource production and manufacturing and transportation processes. Suggest alternate ways that this gas production may have been avoided. (2 marks)
   (c) Evaluate the gases produced during your resource production, manufacturing process and transportation process and suggest how this could be improved. (1 mark)
   (d) Suggest alternate timber sources by researching the contact details of sustainable timber plantations in Australia. Compare the financial and environmental cost of using each type of timber. (1 mark)
   (e) Examine the waste, air and noise pollution and suggest viable green alternatives. (2 marks)
   (f) This must encompass all aspects of the design solution including resource production, manufacturing process and transportation as well as what makes this sustainable. (2 marks)
   (g) Refer to Chapter 14. (2 marks)
   (h) Refer to Chapter 15. (2 marks)
   (i) Refer to Chapter 13. (2 marks)

Chapter 6 p. 69

Key concept question

1. List the major issues in your MDP that you are unsure how to resolve. Then fill in the six-week plan below.

Week 6
Week 5
Week 4
Week 3
Week 2
Week 1

The specific steps needed to carry this work forward include:

a.		Who	When
b.		Who	When
c.		Who	When
d.		Who	When

Sample Preliminary questions

Short-answer questions

1. Cognitive organisers are a more effective aid to comprehension. They are a simplified, visual representation of complex concepts and ideas. (2 marks)

2. Define 'creativity', 'lateral thinking', 'cognition' and 'cognitive organisers'. Discuss how cognitive organisers reframe the problem, increasing the quantity and quality of fresh and potentially valuable ideas. Discuss how creative problem-solving techniques lead to a fresh perspective, using a mental process to create a solution to a problem. Creativity requires innovation and the solution must either have value or solve the stated problem. (5 marks)

Chapter 7 p. 80

Key concept questions

1. High-level responses will relate the cause to effect, making the links evident and providing a why and/or how the design brief contributed to the overall success of the MDP. The design brief should:
   - provide focus and direction
   - reduce the time of the design process
   - make it easier to arrive at an appropriate design solution
   - clarify the need by developing a common understanding
   - be used when testing and evaluating.

2. Appropriate responses will identify the essential qualities of all forms of communication. Students will need to show an understanding of how communication occurs on a series of levels. Suitable forms of communication would include discussion of designer's ideas with the client, showing design sketches and concept boards to the client, and the use of models and/or prototypes.

3. Responses should provide characteristics and features for four design factors. In the final part of the question the student needs to link each design with the appropriateness of each factor. While only four are required, an extended set of factors will help identify all components. Factors include the following.
   - Function: the overriding reason for the cross-trainers is their ability to serve a purpose.
   - Aesthetics: designers will build in features that are visually important or attractive.
   - Finance: all expenditure will be monitored against an expected finance plan.
   - Quality: the company will focus on quality by determining how their product compares to that of their competitors.
   - Short- and long-term environmental consequences: the company will need to evaluate the environmental impacts.
   - Obsolescence: the cross-trainers will become obsolete when they reduce their function due to natural wear, or when the company decides to introduce a new product which supersedes the old.
4. Describe the link between the design idea, evaluation and the vision of success. Explain that before the design idea has reached maturity the designer must undergo a process of evaluation using a list of successful measures or design factors which will determine the success of the final design.
5. Responses should show a clear distinction between the concepts of micro and macro. The easiest approach is to treat each separately and list appropriate solutions. Factors for the trainers may include:

   The micro-environment uses award schemes to develop loyalty with customers. It could develop relationships with sporting teams and advertise the 'healthy image'. Could also include product 'giveaways' and discounted specials.

   The macro-environment could identify health agencies, fitness clubs and political support in promoting health and fitness through exercise.
6. Responses should 'outline' the main features of the marketing mix, presenting each element and relating it back to the expected goals. Start with a brief introduction on the importance of the marketing mix. Factors could include the following.
   - Product: decisions would include aspects such as function, appearance, packaging and warranty.
   - Price: pricing decisions should take into account profit margins and the probable pricing response of competitors.
   - Place: would involve decisions on market coverage, selecting supply chains and logistics of delivery, in-store placement and levels of service.
   - Promotion: involves advertising, loyalty rewards and public relations.
7. Responses should recognise the importance of the elements associated with the end product. Recognition of details includes the outward appearance of the package and the printed plastic bag which, with an emblazoned colour logo, identifies it to the customer and all shoppers in the shopping centre. The warranty and service become part of the product as they support the product through its life cycle. All elements become 'consumer complete' and are absorbed within the detail of the product.
8. Responses would identify the way a company responds to market variations. Students should identify what drives market forces and the ways in which companies attract and keep customers such as by:
   - continually changing to meet the demands of trading forces
   - introducing new products to satisfy customers
   - continuing extensive marketing campaigns
   - obtaining strong support from customers by developing brand loyalty.

Sample Preliminary questions

Objective-response question

1. **D** is the correct option. Before the design idea has reached maturity the designer must undergo a process of evaluation using a list of successful measures or design factors which will determine the success of the final design. By forming a set of standards or criteria the designer can judge the merit, worth and significance of the appropriate design, validating the idea and providing design criteria that will be used at the end to determine the design's success. Option **A** is more about the commercialisation of the innovation followed by marketing formulas, **B** is important in keeping

costs down but this is already realised in the finance plan, and **C** relates to long-term profitability coming from an examination of the design through a life cycle analysis. (1 mark)

Short-answer question

2. In this question you have to 'describe', which means you must provide characteristics and features on what would be the best research methodology to evaluate data about this particular target market. The easiest approach is to identify each research methodology and build on these.

   Primary research entails the use of immediate or up-to-date data in determining the present market. Ways of collecting primary data includes surveys, interviews and the use of focus groups.

   Secondary research is a means of reprocessing and reusing collected information to better a service or product. It is generally used to historically analyse trends. It may be the only data on the subject. (5 marks)

Chapter 8 p. 86

Key concept questions

1. Responses should recognise and name the characteristics and properties of the materials used for designing and directly relate them to their practical application. The key here is to involve as many possibilities as you can, including:
   - material testing
   - comparing
   - experimenting
   - using published technical data or result scales
   - collection of data mining tools, including primary and secondary research techniques
   - engaging the advice of experts.
2. Responses will cover the concept that 'built-in' design complexities need to be avoided at all costs. There are times when a designer tries too hard to leave a 'design imprint', often resulting in a loss of functionality.
3. Responses will recognise the importance of the experiment and its direct association to the reliability of the result. Achieving reliability requires a very tightly controlled environment. Responses should also note that every experiment can be realised so that consistency, predictability and reliability are assured.
4. The easiest approach is to understand the decision-making adopted by designers.
   In developing suitable criteria designers take into account:
   - function
   - performance
   - energy requirements
   - aesthetics
   - finance
   - ethical and legal considerations
   - properties of materials and their relation to the suitability of a production technique.
5. Responses will identify a range of strategies used to make decisions based on the understanding of the physical properties and working characteristics of materials chosen. A useful approach is to plan out each process in a sequential order as below.
   - Concept sketching done quickly.
   - Design sketching produces more detailed ideas.
   - A variety of 3-D modelling media such as cardboard, clay or foam can be used, as can rapid prototyping and 3-D printing.
6. Responses will identify that each material has its own method of processing and therefore its own set of safety considerations. Students need to treat each separately. Factors include the following.
   - Properties: having knowledge of the properties of different materials, for example, cardboard, wood, metals, glass, fabrics and polymers and the processes that can be safely applied to them.
   - Production techniques and equipment: considering the safe and competent use of production techniques and equipment and recognising that training and supervision is required.
   - Risk assessment: applying a risk assessment to each process will provide information on how to care for and handle equipment and materials safely and manage the reporting of faults.

Sample Preliminary questions

Objective-response question

1. **D** is the correct option. Designers should maintain control over the selection of materials by controlling the tools and processes used.

This will not only maintain quality but also provide control over safety throughout each process. For option **A**, while results are valuable they will not recognise the suitability for a particular application. With **B**, location is an important consideration when selecting resources but a designer would not select these based on location alone. For **C**, even if the designer displays strong skills the technique will be defined by the tools used on the preferred material. (1 mark)

Short-answer question

1. In this response students must provide a 'why and how' of how varied materials are applied by designers in planning out their design decisions. Begin by recognising that a range of materials, tools and techniques are used by the designer and carried out in unison. Identify how, as the designer moves to more complex thought processes, so does the complexity of the design strategy. (5 marks)

Chapter 9 pp. 93–94

Key concept questions

1. Responses should clearly relate the importance of evaluation in developing a structure of control within the management of the project. Students should indicate how:
   - moving through the design process too quickly may result in some important aspects failing to be recognised
   - both finance and action/time plans should be prepared
   - if the planning is too tight then some element in the process may be required to be reduced or scaled back.
2. The question relates to the importance of evaluating all aspects of decision-making. Responses should include the idea of continuous evaluation as it will:
   - prevent designers from losing focus
   - reduce the chance of failure
   - improve competitive strength.
3. Functional criteria is based on the working operation of the product, such as how it caters for its users, its features, and product security such as warranties and ease of service. Aesthetic criteria evaluates aesthetic success based on the appearance (lines, style, texture, colour) of the product.
4. This question involves analysis of the criteria on a specific product. A recommended approach is to describe each issue separately. The following list presents a variety of design considerations that can be used in the analysis.

   Function as criteria:
   - caters for a variety of user groups
   - SIM slot capability
   - provides a variety of features
   - data volume
   - has clear on-screen instructions
   - simple pathways to features
   - versatile language base
   - user-friendly control system
   - strong warranty
   - acceptable size.

   Aesthetics as criteria:
   - variety of colours
   - styles which appeal to the market
   - good lines and an efficient shape
   - textural to hold
   - consistent page display
   - text characters are appealing
   - strong use of graphic display.
5. Responses should provide the characteristics and features of particular test methodologies. Students need to be aware of the link between the appropriateness of the test in supporting the success of the product. Appropriate forms of testing would include:
   - fishing rod—controlled field testing of the product
   - energy drink—studies that examine consumer use and attitudes
   - toothbrush—surveys that focus on product satisfaction
   - remote control device—controlled field testing of the product
   - socks—studies that examine consumer use of the design.
6. In this response students are required to make a judgement of value. They should think like a designer by 'stepping inside the shoes of the designer'. In this way all design considerations are explored including:

- effects on individuals, such as safety
- ease of use
- designed obsolescence
- toxicity of the material, paint or dye
- small parts which may present a choking hazard
- unique design for minority groups
- sustainability of resources and the disposal of end products
- cradle to cradle solutions.

Sample Preliminary question

Extended-response question

1. Responses need to 'analyse', which means students must identify components and the relationship between them and the implications of this. They should identify the design and the responsibility placed on the designer. They should describe the design and the problems and conditions associated with poor designing. They should draw out the idea of safeguards and measures used to protect continued use of the design. Finally, they should explain the importance of product labelling and the responsibility on consumers to read these labels.

Marking guidelines criteria	Marks
■ Identifies the relationship that is developed between the concept of the child's toy and the safety hazard of choking on small parts ■ Identifies the tendency of small children to place objects in their mouth as part of a tactile learning experience ■ Draws out and relates implications between the idea of a breakage while the child is playing with the toy ■ Draws out the idea of safeguards and measures that are there to protect continued use of that design. Explains the importance of product labelling and the responsibility of consumers to read this information ■ Response is well supported by returning to the stimulus image as an example	12–15
■ Identifies the relationship that is developed between the concept of the child's toy and the safety hazard of choking on small parts ■ Identifies the tendency of small children to place objects in their mouth as part of a tactile learning experience ■ Draws out and relates implications between the idea of a breakage while the child is playing with the toy ■ Identifies a safeguard(s) and measure(s) that are there to protect continued use of the design ■ Response may be supported by returning to the stimulus image as an example	10–11
■ Draws out and relates implications between the idea of a breakage while the child is playing with the toy ■ Draws out the idea of safeguards and measures that are there to protect continued use of the design	9
■ Explains the relationship between the breaking of a small part and the possibility of choking	7–8
■ Discusses features of the choking hazard and its impact on the child	5–6
■ Describes the choking hazard and its impact on the child	3–4
■ Outlines the choking hazard OR an impact on the child	2
■ Identifies the choking hazard OR identifies an impact on the child	1–2

Chapter 10 pp. 102–103

Key concept questions

1. There are three forms of communication (verbal, nonverbal and graphic) and four types (intrapersonal, interpersonal, public and mass communication).
2. A prototype or model is used as part of the design process to allow exploration of design alternatives, to test theories and to confirm performance prior to starting production of a new product. An interactive series of prototypes will be designed, constructed and tested as the final design emerges and is prepared for production. A common strategy is to design, test, evaluate and then modify the design based on an analysis of the prototype.
3. Communication involves the visualisation of clear design ideas and solutions. The solution must involve a common understanding.

Sample Preliminary questions

Objective-response questions

1. Option **A** is the only answer that mentions oral communication, which is the basis of the question.
2. **D** is the correct option. Communication occurs using a range of styles, however, verbal is the most commonly used interpersonal communication method.
3. Option **B** is correct because the term 'element' refers to the basic components used in this process.
4. Option **C** is correct because while all components are necessary to convey a clear message it is body language that makes it memorable.
5. Option **B** is correct because when presenting design solutions to a client it is important to use the rule of three because people usually only remember three things. (1 mark each)

Short-answer questions

6. Communication is a type of communicology and is concerned with transferring information from one entity to another. It is the 'imparting or interchange of thoughts, opinions or information by speech, writing or signs'. Collaborative communication involves a team of people working together and communicating in a range of ways. (2 marks
7. Common barriers to successful communication include message overload (when a person receives too many messages at the same time) and message complexity (refers to the degree of difficulty in decoding the message). To communicate successfully use simple, to the point messages delivered individually. (4 marks)

Extended-response questions

8. Successful presentation techniques include:
   - the use of visual aids
   - keeping oral presentations to under twenty minutes
   - using the rule of three
   - rehearsing the presentation
   - using narratives, stories and anecdotes to help illustrate points
   - not putting notes up on the screen
   - knowing what slide is coming next
   - having a back-up plan in case the technology fails
   - evaluating the presentation room by arriving early. (15 marks)
9. (a) A passive person would react by ignoring them and perhaps smiling at them.

   An aggressive person would start screaming at the others and bullying them into submission.

   A passive-aggressive person would react by smiling and laughing with them and then sneaking out to tell the boss.

   An assertive person would react by taking aside the main person and calmly explaining that they need to complete their work and quietly insisting that they allow this to occur. (7 marks)

   (b) A passive person would react by not saying anything.

   An aggressive person would react by pushing the person out of the way, snatching the laptop back and walking out.

   A passive-aggressive person would react by not saying anything to the friend but complaining to the teacher.

   An assertive person would react by firmly but in a positive way explaining to their friend what the situation was and requesting the immediate return of the computer. (8 marks)

Chapter 11 pp. 108–109

Key concept questions

1. Factors include appropriateness of the design solution, needs, function, aesthetics, finance, ergonomics, occupational health and safety, quality, short- and long-term environmental consequences, obsolescence and life cycle analysis.
2. Obsolescence is the state of being which occurs when a person, object, or service is no longer wanted even though it may still be in good working order. The types are outlined below.

   Technical obsolescence, which may occur when a new product or technology supersedes the old, and it becomes preferred to utilise the new technology.

   Functional obsolescence: particular items may become functionally obsolete when they do not function in the manner that they did when they were created.

Planned obsolescence: sometimes marketers deliberately introduce obsolescence into their product strategy with the objective of generating long-term sales volume by reducing the time between repeat purchases.

3. Organisational factors include characteristics of task, work flow, ergonomics and manufacturing and work practices. Environmental, social and cultural factors include personal values, cultural beliefs and individual and community needs.

4. Life cycle assessment is used to evaluate the environmental impact associated with a product, process or activity by identifying and assessing the energy and materials used. The assessment includes the entire life cycle of the product: extracting and processing raw materials; manufacturing, transportation and distribution; use and reuse; maintenance; recycling and final disposal. Responsible designers consider how every idea, material, tool and technique used will impact on the natural world and then attempt to minimise this impact.

Sample HSC questions

Extended-response questions

1. Factors which need to be discussed include:
   - appropriateness of the design solution
   - needs
   - function
   - aesthetics
   - finance
   - ergonomics
   - occupational health and safety
   - quality
   - short- and long-term environmental consequences
   - obsolescence
   - life cycle analysis.

2. The 15 possible marks for this question would be allocated along the lines of the material below.

   For 3 marks: description of MDP.

   For 6 marks: an explanation is provided of how responsible environmental design has been considered including:
   - life cycle analysis
   - materials and processing of materials
   - tools
   - techniques and manufacturing process
   - transportation
   - carbon footprint
   - cradle to cradle.

   For 6 marks students need to discuss the following factors:
   - appropriateness of the design solution
   - needs
   - function
   - aesthetics
   - finance
   - ergonomics
   - occupational health and safety
   - quality
   - short- and long-term environmental consequences
   - obsolescence
   - life cycle analysis.

Chapter 12 p. 117

Key concept questions

1. Responses need to discuss areas of design thinking such as:
   - a focus on customers/users
   - redesigning
   - finding alternatives
   - reclaiming design
   - practical applications to the sustainable design theory
   - a wide range of influences used to draw inspiration and solutions
   - design that incorporates inspiration and emotion
   - the use of intelligent humour in the design.

2. Design can be a very personal thing and needs to provide a connection between the design and the person. The end user needs to generate a positive feeling from an object through touch and sight. Many designs gain inspiration from the natural world. Designers are careful in maintaining an emotional connection with the object even when working with the most recent techniques and materials.

3. No longer do companies rely upon design staff to imagine the design. By employing a well-known designer the company will be provided with some current design thinking. The designer is free to work without imposed constraints and is not

restricted by office pressures. The company can also later highlight the use of the designer in their marketing campaign.

4. To list the main processes of a designer is firstly to understand the materials, tools and technologies used. Marcel Wanders uses high technology in a very traditional way. An example is his Knotted Chair where he uses carbon fibre and resin to complete a rigid form. He thinks cleverly as he takes current technology and mixes it with good form and high design.

   To describe the designer as a 'design thinker' you need to understand their design philosophy. Marcel Wanders never loses the value that is placed on human emotion. His designs appeal to human emotion by providing fun and excitement. An example is the lamp that requires the user to blow out an electric candle.

   To identify the factors that contribute to the success of a designer you need to look at their inspirational motivation. Marcel Wanders is concerned about making his design products meaningful so that they will have value and never be thrown out. He sees their life cycle as infinite as each product gets handed down from person to person thus recreating the whole concept of product reuse.

Sample HSC questions

Extended-response question

1. (a) In this question you have to 'identify', which means to recognise and name some of the practical applications used as common principles in sustainable design. Due to the importance of a strong environmental future there are growing trends from all designers to look at the possibilities of sustainable design. Referred to as 'green design', 'eco-design' or 'design for environment', designers select materials and manufacturing processes which help reduce the carbon footprint. To obtain five marks you need to discuss the practical applications which are identified as sustainable design. For three to four marks some elements of sustainable design would be identified while a limited knowledge of how sustainable design can be used in practical applications would gain one or two marks.

   (b) In this question you have to 'describe', which means to provide the characteristics and features of the practical applications that you have used in your MDP. The aim for a Design and Technology student is to produce products, systems and environments that reduce the use of non-renewable resources. Some of the practical applications of sustainable design theory involve:
   - choosing low-impact materials such as non-toxic, sustainably produced or recycled materials which require little energy to process
   - energy efficiency: using manufacturing processes and producing products which require less energy
   - quality and durability: longer-lasting and better functioning products will need to be replaced less frequently, reducing the impacts of producing replacements
   - design for reuse and recycling: using cradle to cradle methodology.

Chapter 13 pp. 125–126

Key concept questions

1. Responses will recognise that customer preferences determine the styles and designs that are developed through to mass production. Mass production efficiency, or mass customisation, has become a major trend in industry. To meet an increasingly sophisticated customer demand, companies must be able to:
   - provide a higher degree of product customisation and flexible production systems
   - have the ability to cope with extended product ranges
   - adopt the concept known as 'design by customers'
   - communicate to customers what the company can offer
   - find out customer needs
   - assist customers in making choices
   - make final product agreements.

2. Students need to cover:
   - the idea that when companies bring out a new product line or breakthrough invention they tend to leapfrog their competitors
   - the relationship of the consumer to the newly developed or improved feature

- most hype surrounding new products tends to appeal to a consumer liking for fads and gadgets and it is this emotion that companies like Apple exploit
- styles of marketing that evoke excitement with every new product launch.

3. Students need to recognise how the product is taken from the initial conception to its mass production. The elements that impact on a flexible production system include:
   - a shorter product life cycle
   - rapid prototyping technologies
   - the emergence of sophisticated software environments, systems and management processes
   - clear product supply chains requiring high-end transportation logistics
   - choice of supplier involving personal and moral commitments
   - suppliers being involved in the design and development of new systems
   - an ability to adapt products to market and technology changes.
4. A Venn diagram organises the ideas into an order or organisation. It immediately identifies the relationships between each element.

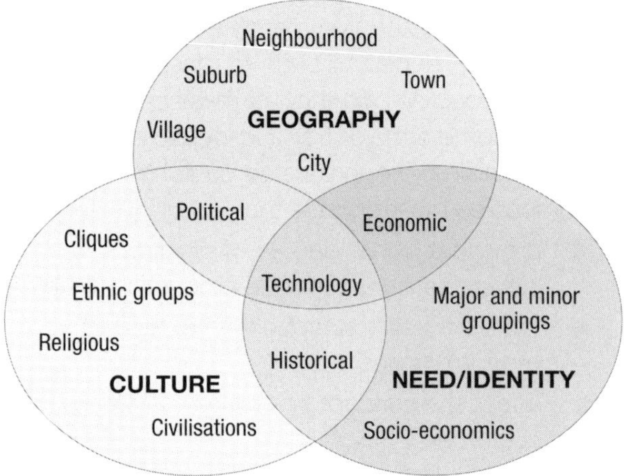

5. To 'describe', students need to provide information on the characteristics and features of the roles of these 'watch dog' groups. These groups provide protection to the consumer by outlining the responsibilities of companies and manufacturers.

   The Australian Competition Consumer Commission (ACCC) controls any unfair competition practices in Australia while the Foreign Investment Review Board controls foreign involvement. Standards Australia is a regulatory body that controls innovation through a series of standards that the innovation must meet before it can proceed to commercialisation.
6. To keep costs to consumers down and maintain a healthy profit margin, some companies move their manufacturing base to places with cheaper labour forces. This concept of taking the company 'offshore' means that manufacturing costs are dramatically reduced due to significantly lower wages. This can have a crippling effect in Australia, resulting in job losses and a reduced amount of national resources being used.
7. One of the factors facing the marketplace is the evolution of design trends. These might start as a fad or a craze and then take on some real significance. Customers' preferences are the motivation behind the designing and producing and are exhibited in the following areas:
   - social pressure
   - patriotism
   - materialism
   - multiculturalism
   - social class
   - global issues
   - political issues.

Sample HSC questions

Objective-response questions

1. **C** is the correct option. In today's work environment it is necessary to be multi-skilled, with most workers being required to have technical skills that enable them to operate data entry, word document and web-based applications.
2. **B** is the correct option. An efficient plan towards new technology should ensure that the older technology is reduced through a series of recycling processes.
3. **A** is the correct option. The size of the market will influence the methods of design and production. The larger the market, the greater the industry required in producing. The designing will be directly related to the production method(s).

(1 mark each)

Chapter 14 p. 134

Key concept questions

1. Intellectual property (IP) represents the property of your mind. It can be an invention, trademark, design or the practical application of a good idea.
2. A patent grants protection to any device, substance, method or process which is new, inventive and useful. It is legally enforceable and gives the owner the exclusive right to commercially exploit the invention for the life of the patent.
3. Design ethics is when designers consider the ecological and psychological consequences of their design.
4. Reuse: the existing product is reused for another function; renew: the resource or materials used are renewed so the original supply is not exhausted; recycle: the product is recycled (adapted) into a new product.
5. Design impact measures the total earth ecological footprint. This means that material extraction, processing, manufacturing, transportation, packaging and marketing must be analysed for environmental impact. This is done through carrying out a life cycle assessment. Sustainable design standards are becoming increasingly available to assist with this analysis.

Sample HSC questions

Extended-response questions

1. Students need to define 'ethical design' and 'environmental design'. They must discuss how designs reflect moral and ethical roles in society and that there is a responsibility that comes with this. Design trends are influenced by culture, innovative technology and the economy. Responses should include information on the following:
   - corporate and commercial designers and consumers
   - consumption, production and waste
   - environmental issues
   - links between consumerism and the environment
   - 'true cost' to consider the ecological and psychological consequences of design
   - design code of ethics.

Ethical design involves applying talents to serve a social or cultural purpose, or fill a need in society. If every designer takes ethical issues to heart and strives to uphold their morals, design can become an industry defined by honest, important and positive communication and social change. (15 marks)

2. Students need to define 'sustainable design', 'sustainability' and 'appropriate technology'. They must describe how the most appropriate technology may not be the most sustainable one, and how some sustainable technologies have high cost or maintenance requirements. They should also explain the principles of sustainable design, including:
   - low-impact materials
   - energy efficiency
   - quality and durability
   - design for reuse and recycling
   - design impact measures
   - sustainable design standards
   - biomimicry
   - service substitution
   - renewability
   - healthy buildings. (15 marks)
3. Define intellectual property (IP) and patents. Describe how patents protect IP and how obtaining one adds value to your invention and business in the following ways.
   - You can benefit from the financial rewards.
   - You can bring patent infringement proceedings against any unauthorised user of the invention.
   - Having a patented invention will assist in forming relationships with potential partners, venture capitalists and licensees.
   - A patent protects the idea during the time and investment that you have to put into research and development.
   - An Australian patent provides protection only within Australia. (15 marks)

Chapter 15 p. 142

Key concept questions

1. To 'explain' is to relate cause and effect. In relation to this, an innovation is:
   - the process of discovering or creating something new
   - often based on an existing invention that is modified or improved to suit a changing need.

2. Students should identify and list all factors as below.
   - Timing: the timing of an innovation generally occurs when a company or consumer perceives a need.
   - Available and emerging technologies: these can provide the stimulus to initiate the innovation process.
   - Historical, cultural and political factors: these form the base for society's basic values, perceptions, preferences and behaviours.
   - Economic and legal factors: these involve laws, government agencies and pressure groups.
   - Marketing strategies: the process of planning and executing the conception, pricing, promotion and distribution of ideas, goods, and services. Successful marketing strategies convince consumers to buy their product.
3. Students need to cover how agencies are set up to provide support via financial grants, processes, regulations, laws and review procedures that organisations need to follow. Agencies work through a number of distinct stages, beginning with the formulation of a novel idea/concept and, through a series of stages, ending in the successful launching and marketing of a new or improved product in the marketplace. Within these agencies are divisions or sections where individuals participate as technical experts in research and development, marketing, management, finance, legal, and so on, in addition to outside consultants, suppliers, component manufacturers/service providers, business partners and lead users.
4. The four major design centres are New York (USA), Paris (France), Milan (Italy) and London (England).

5. Entrepreneurs identify market opportunities and exploit them by turning ideas into marketable successes. Entrepreneurs are willing to accept a high level of personal, professional or financial risk to pursue an opportunity.
6. All successful entrepreneurs have the following qualities:
   - an inner drive to succeed
   - a strong belief in themselves
   - constant search for new ideas and innovation
   - an openness to change
   - competitive by nature
   - highly motivated and energetic
   - accepting of constructive criticism.

Sample HSC questions

Extended-response question

1. (a) Rip Curl's H-Bomb is an example of a successful innovation designed specifically to meet a target market of water sports enthusiasts.

   (i) The main factor of the innovation is it allows surf enthusiasts to participate in water sports during winter or in areas of extreme cold. (2 marks)

   For 2 marks: linking the design idea to its function.

   For 1 mark: recognition of the idea but no strong link to design and function.

   (ii) In this question you have to 'identify', which means you must recognise and name the functions of the wetsuit. The wetsuit has three different temperature levels and the surfer can regulate their heat for up to two and a half hours. The heating elements are non-metal based and use carbon fibre technology to conduct the heat, taking advantage of the non-corrosive material. A minimal electromagnetic field is generated from the technology. (3 marks)

   For 3 marks: each function of the wetsuit is identified and named.

   For 2 marks: the functions are either identified or named.

   For 1 mark: provides a general listing or identification of the functions.

(b) In this question you have to 'outline', which means you must sketch in general terms the research that would have been undertaken. Research would involve testing of the lithium batteries and their ability to drive an element to a desired temperature. Research could also involve switches to turn the unit on and vary the temperature settings. There would also have been extensive testing done in marine environments and cold water conditions. (3 marks)

For 3 marks: methods of research and testing are outlined and link the test to the application.

For 2 marks: methods of research and testing are identified with limited linking of the test to the application.

For 1 mark: recognition of the applications and/or identification of a test.

(c) (i) In this question you have to 'explain', which means you must relate the cause and affect. The main reason for protection of ideas is to prevent others copying them. Designers and companies should be the ones who benefit from their innovations. (2 marks)

For 2 marks: identifying the relationship between the reason for protection and the benefits.

For 1 mark: a reason and/or benefit identified.

(ii) In this question you have to 'list' the strategies. You can draw from any of the following:

- patents for new or improved products or processes
- trademarks
- aspects of packaging that distinguish the goods and services of one trader from another
- copyright for original material in literary, artistic, dramatic or musical works, films, multimedia and computer programs
- circuit layout rights
- plant breeder's rights
- confidentiality/trade secrets. (5 marks)

For 5–4 marks: three strategies listed with an explanation of how they are used in designing.

For 3–2 marks: strategies listed with a limited explanation of how they are used in designing.

For 1 mark: strategy listed.

Chapter 16 p. 151

Key concept questions

1. Strategies for innovation include:
   - addressing old problems in new ways
   - finding new ways to use technology
   - using sustainable design
   - investigating new styles
   - addressing new user groups
   - the application of miniaturisation and technology reduction
   - the creation of multifunctional products.
2. Designers are able to tap into various stimuli and use them as their source of inspiration. Stimuli can include music, aromas, observations and photographs among many other forms.
3. Designing with a 'high ceiling and low walls' is about:
   - exploring all ideas
   - keeping thoughts flexible
   - supporting collaboration and new ideas.
4. Any functions that remain hidden from the user's view are known as 'black box'. At most times the consumer is not affected by the processes or ideas surrounding the design and will be more concerned with presentation, function and aesthetics.
5. Methods include:
   - defining tasks by taking risks and thinking creatively
   - identifying new customers with unmet needs
   - targeting existing customers
   - being goal driven
   - establishing collaborative teams
   - effective communication
   - maintaining continuous improvement via TQM
   - maintaining a sustained process of investment.

Sample HSC questions

Objective-response questions

1. **D** is the correct option. Innovation is about discovering new methods and making changes and technology is one way innovation can proceed.
2. **B** is the correct option. Almost by definition, creative work means that the final design is not necessarily known at the outset so designers must be encouraged to explore all ideas.
3. **C** is the correct option. Most successful designers seek to understand their users in order to design products well-matched to their needs.

Chapter 17 p. 157

Key concept questions

1. Labels are:
   - the emerging technology
   - the innovation
   - designs and design practice
   - factors which may impact on successful innovation
   - entrepreneurial activity
   - the impact of emerging technologies
   - the impact on Australian society
   - historical and cultural influences
   - ethical and environmental issues
   - creativity.
2. You need enough information on each of the above topics to write a detailed summary or a HSC essay as an exam question could be asked on any of these topics.
3. The headings are listed below.

   Title page

   Contents page

   Abstract

   Acknowledgements

   Main body of the report
   - Introduction
   - Emerging technology
   - Innovation
   - Designs and design practice
   - Factors which may impact on successful innovation
   - Entrepreneurial activity
   - The impact of emerging technologies
   - The impact on Australian society
   - Historical and cultural influences
   - Ethical and environmental issues
   - Creativity
   - Conclusion

   Reference list

   Appendix

   List of figures or details
4. You can start over using a different innovation and emerging technology or, alternatively, more thoroughly research the topic using both primary and secondary research.
5. Emerging technologies are new and/or improved materials, tools, techniques, products, systems or environments that allow new innovations to occur. Innovations are usually linked to emerging technologies as people learn to use the technology in new ways.

Chapter 18 p. 162

Key concept questions

1. Optalert driving glasses are designed for people who drive for extended periods. This design combines two existing technologies to form a new innovation. The first technology was used to test and monitor the fatigue. The second technology involved the use of a pair of glasses with lenses that are graded to protect the eyes from harmful rays. The designer recognised the practical nature of this innovation and placed the two together.
2. Responses should name the following factors:
   - timing
   - available and emerging technologies
   - cultural factors
   - political factors
   - economic factors
   - marketing strategies
   - patents office, Standards Australia
   - legal factors.
3. The Optalert technology is addressing a critical industry need by providing an effective method of reducing accidents caused by drowsiness. The entrepreneurial activity is to develop the product through to commercialised success. To aid in the success of Optalert the following processes were adopted:
   - Australian designed and financed product
   - identification of the need
   - research and testing
   - development of marketing and business plans
   - support from the NRMA and the government.
4. Responses will provide characteristics and features of the innovation's technology including the following.
   - The technologically advanced sensors detect movements in the eyes and eyelids.
   - These movements are distinguished by Optalert with variables calculated.
   - It utilises the use of a driver alert warning system.
   - It uses back to trucking company base monitoring.

Sample HSC questions

Short-answer questions

1. In this question you have to 'analyse', which means you must identify the impact by recognising the relationship with Australian society and drawing out the implications. (3 marks)

   For 3 marks: impact of innovation is identified and its relationship on Australian society drawn out.

   For 2–3 marks: impact of innovation is identified with limited relationship to Australian society described.

   For 1–2 marks: some aspects of the innovation described.

2. In this question it is best if you identify all the 'way points' throughout the process and place them down in chronological order. (2 marks)

   Need → Research the need → Research available and emerging technology → Testing and design development → Problem investigation and solving → Consideration of environmental, social and cultural impacts → Marketing → Business plan → Consideration of factors impacting on success → Financial support and government grants from agencies and organisations.

   For 2 marks: a detailed timeline is drawn indicating the historical development of the innovation

   For 1 mark: a timeline is drawn but the historical development of the innovation requires greater detail

3. In this question the verb used is 'describe'. This means that you must identify and provide the characteristics and cultural features of your innovation. (2 marks)

   For 2 marks: characteristics and features of the impact of the cultural features are described.

   For 1 mark: characteristics and/or features of the impact of the cultural features are identified.

4. In this question you have to 'outline', which means you must sketch in general terms the social and ethical issues associated with your innovation. Begin by developing the idea of design trends and work through this. (3 marks)

   For 3 marks: social and ethical issues outlined and described in specific terms.

   For 2 marks: social or ethical issues outlined and described in specific terms.

   For 1 mark: social or ethical issues outlined in general terms.

5. In this question you have to 'evaluate', which means you must make a judgement and determine a value based on the environmental impact of your innovation. Develop the idea of technology-specific environmental standards in developing clean and sustainable manufacturing technologies. (5 marks)

   For 5 marks: a strong judgement with value based on the environmental impact closely related to the innovation.

   For 3–4 marks: a judgement with value based on the environmental impact loosely related to the innovation.

   For 1–2 marks: a judgement with value based unrelated to the innovation.

Chapter 19 p. 169

Sample HSC questions

Extended-response questions

1. The innovation is the fast suit. The concept revolves around a swimsuit which can cover you without adding to your drag and in addition can reduce your drag to help you swim more quickly without using more energy or oxygen.

   Material: the fast suit is a sheet of woven spandex, with smooth panels sealed onto the torso, legs and rear. The smoothness comes from the materials which act like kitchen cling-wrap.

   Construction: constructed from three pieces of welded fabric to minimise seams.

   Physiological impact: the suit has a core stabiliser. This is a girdle where the material is doubled. Swimmers use their core muscles to find the posture in the water that helps them float the best.

   Emerging technology includes the following.

   - Weight: the swimsuit is very light. This specifically designed spandex weighs 100 grams per square metre, four times less than standard swimsuit material.
   - Waterproof: the spandex is coated with a water-repellent substance.
   - Seams: the swimsuit has a low profile zip and welded and bonded smooth seams.

Physiology: a hydro compression system produces a hydrodynamic compressed shape across the whole body, allowing a range of movement. It is energy efficient.

Effect: in simulated race conditions swimmers swam four per cent faster and used five per cent less oxygen. (15 marks)

2. The factors which contributed to successful innovation are listed below.

   Consumer:
   - They are proven performance enhancers for swimmers, allowing athletes to swim at their personal best.
   - They are water-resistant and reduce drag and muscle fatigue by up to four per cent.
   - They are lightweight.
   - They are slowly becoming available to all consumers.

   Manufacturer:
   - All three fast suits have proven to be extremely successful, with over forty world records broken by wearers.
   - Due to the quick obsolescence of the suits there is constant demand and continual manufacturing.

   Environment:
   - FINA is constantly reversing decisions on the legality of fast suits for all global competitive events. This is due to the fast suit technologies being further developed and continually evolving.
   - Current record holders who have worn the fast suits will retain their world records even if the suits are banned in the future.
   - Because the suits are not mass produced this will not impact on issues such as manufacturing pollution or extreme wastage.

   The factors which hindered successful innovation are listed below.

   Consumer:
   - Fast suits are very expensive, ranging from $400–$600, and are not easily available for general purchase.
   - They take up to thirty minutes to put on.
   - Because of the ultra-thin, lightweight stretch fabric the suits can easily tear.
   - Suits stretch after several wears and therefore become ineffective.

   Manufacturer:
   - Due to the specific design pattern of suit panels there is an increase in fabric waste making the suits costly for Speedo to produce and thereby raising the price.
   - Controversy over performance-enhancing suits taints the Speedo name and reputation.

   Environment:
   - Dramatic fabric waste.
   - Because the suits can be easily torn or stretched after several wears they quickly become obsolete. (15 marks)

3. The competitive swimwear industry continues to grow: almost all swimwear companies make some sort of fast suit and market them in different ways. Adidas sent every swimmer who qualified for the US Olympic trials a full body suit though marketing ploys as generous as this are rare. The producing companies will continue with their desire to corner the market by continuing with research and development of fast suits. (15 marks)

4. The emerging technologies surrounding fast suits include:
   - reducing drag
   - use of water-repellent fabrics
   - muscle compression technology
   - use of bonded and welded fabric seams
   - replicating the dermal denticals of sharkskin. (15 marks)

5. Athletics has a positive impact on society: physically, recreationally, emotionally and psychologically. The idea behind modern athletics is that an individual can achieve on his or her own. By implementing a fast suit the line between acceptable and unacceptable 'help' has been blurred.

   Australians, both as competitors and spectators, all have an opinion on the fast suits and whether they are simply sporting costumes/equipment just like a bobsled is in the winter Olympics or are they a form of cheating. With constant evolution of materials one must consider when, if ever, the design of athletic equipment is complete. Access to this equipment by amateur athletes will cause swimming authorities more problems as the desire to wear them becomes more widespread.

6. The timeline could include some of the information below.

- 1890s: women raced in stockings, bloomers and short-sleeved dresses.
- 1907: movie star and athlete Annette Kellerman arrested for wearing a one-piece suit to the beach.
- 1940s: Speedo designer Gloria Smythe developed Aboriginal-inspired imagery and prints.
- 1970s: male swimmers wore briefs the size of dinner napkins.
- 1978: Speedo applied Aboriginal-inspired motifs to its competitive swimwear though the Mabo case later led to a reconsideration of the unauthorised use of Indigenous motifs.
- 1980s: most of the world's top competitive swimmers were competing in Speedos.
- 2010: swimsuits are again approaching full-body coverage. (15 marks)

7. Some of the impacts are listed below.

   Cultural:
   - Branches of some religions do not allow followers to show any part or shape of their body in public.
   - Fast suits cover the body but its shape is still displayed.
   - Most followers of these religions choose not to participate in swimming as a sport.

   Costs:
   - Many developing countries cannot afford the costs of these swimsuits.
   - There are high costs for development, research, fitting and materials.
   - The cost can lead to athletes being disadvantaged if they cannot afford the suits.

   Fairness:
   - Does the use of the fast suits as a piece of athletic equipment remove the even playing field that the Olympics strives to achieve?
8. Some of the social and ethical issues surrounding the use of fast suits include the following.
   - Do they help athletes swim faster than their personal best would be otherwise?
   - Do they provide an advantage to wealthier countries?
   - Do they remove the even playing field for all athletes?
   - The debate centres on whether the suits do not enhance the swimmer's performance beyond their ability but simply allows them to reach it without the interference of a poorly manufactured swimsuit.
9. Some of the environmental impacts are outlined below.
   - 'Custom made' means that they cannot be mass produced, resulting in minimised manufacturing pollution and wastage.
   - Fabric waste is high as each panel is cut and moulded specifically.
   - The fabric is made from fossil fuel derivatives.
   - Made from non-renewable resources and leaves an undesirable carbon footprint.
   - Obsolescence is a major issue.

Chapter 20 p. 179

Key concept questions

1. The use of quantitative data helps reinforce the problem and hence clarify the need.
2. The criteria for success determines what the designer needs to do to achieve a successful design solution. This criteria is determined before the solution is developed and the solution must meet every criteria listed. The criteria is developed at the planning stage of the project and the criteria to evaluate success works with the proposal in creating a 'perfect' image of what needs to be achieved.
3. The time and action plan provides the necessary management features to ensure all tasks are completed on time.
4. Students should select only those factors that relate appropriately to their design. These specific design factors directly determine practical and obtainable outcomes.
5. These include:
   - always using a variety of methodologies
   - using the correct communication techniques. to suit the application
   - using multimedia, scripting and/or storyboards.
   - using concept boards, survey results, prototypes, drawings and sketches.
6. Ongoing evaluation provides the total picture from start to finish while the final evaluation is judged against the pre-existing criteria nominated in the project proposal.

Sample HSC questions

Objective-response questions

1. The correct option is **A**. The plan provides the necessary management features to ensure everything is done on time.
2. **D** is the correct option as it contains all the underlying elements of the design process.

(1 mark each)

Short-answer question

3. The best responses will provide characteristics and features of the factors. The factors include the following.
   - Appropriateness of the design solution: raised track above the congestion.
   - Needs: intercity transportation.
   - Function: provides link for tourists etc.
   - Aesthetics: strong styling, good colour and advertising signage.
   - Finance: initial cost offset by fares.
   - Ergonomics: seating available yet standing room for short trips.
   - Occupational health and safety: train meets the platform, operation is supervised.
   - Quality: able to withstand the stresses of use.
   - Short- and long-term environmental consequences: helps to keep cars away from the city.
   - Obsolescence: requires a continued program of maintenance and replacement.
   - Life cycle analysis: many components are able to be reused. (2 marks)

For 2 marks: two design factors identified with explanation of characteristics and features.

For 1 mark: design factors only identified.

24—Answers to Sample HSC Examinations

Section I: Objective-response questions

1. **B** is the correct option. A major part of the collaborative design process is that all members participate in idea development activities and work towards a common goal.
2. **C** is the correct option. A product developed for global acceptance must have its design standardised with minimal variation.
3. **A** is the correct option. There is a definite need here for designers to solve the problem using appropriate technology.
4. **B** is the correct option. The advent of new technology was the stimulus for the innovation of the product. It was most certainly placed on the market with upgrades, new functions or new shapes and colours.
5. **B** is the correct option. Owners of IP rights have exclusive rights relating to their particular property. If someone else attempts to exercise these rights without permission they infringe the IP rights of the owner.
6. **A** is the correction option. Obtaining sustainable power is the factor which presents a total saving to the environment.
7. **D** is the correct option. While the other factors are important they exist as infrastructure to the swimming complex and are not seen as ethical design.
8. **C** is the correct option. The question refers to the functional criteria.
9. **A** is the correct option.
10. **A** is the correct option.

Section II

Short-answer responses

Question 11

(a) (i) The incorporation of a clock radio with digital technology allowing for digital channels, interconnectivity to devices and the supported use as a digital photo frame makes this a very functional device for the home.

(ii) Internet interconnectivity of the digital radio allows users access to social networking sites such as Facebook and Twitter. (2 marks)

(b) Innovation is repetitive in the way it continuously refines and improves as a process throughout the product development, manufacturing, marketing, distribution and product enhancement stages. It is this adhering to product improvement that establishes the advantage over competing products or processes. An innovation can be either:

- a completely new idea or concept
- an improvement of an existing design
- a new design that has had a section of the design improved through problem-solving.

(2 marks)

(c)

Cost	A very expensive product to buy. Generally most new technology is expensive in its initial release. Once the product grows in popularity and consumer faith the manufacturing will be more cost effective, thus bringing the price down.
Function	A very multifunctional product. Allows the user to move through a variety of modes or tasks with efficiency. Provides a large front screen controlled via remote control.
Aesthetics	Very contemporary styling involving the use of elliptical or oval lines. The surface is quite smooth and features blocks of colour. The stand makes the radio a 'feature' as it sits above the table. (3 marks)

(d) (i) Timing: the innovation should succeed as social networking has become a global phenomenon and people are looking for further gadgets to enhance their opportunity to access these networking sites.

(ii) Available and emerging technologies: the technological discovery and invention of the digital radio has provided the stimulus to initiate the innovation process. (2 marks)

(e) The Australian Government recognises that the lack of regulation on social networking sites has spawned a variety of anti-social and criminal activities, ranging from cyber bullying to phishing and theft. By enforcing regulations the government aims to control the situation. (2 marks)

(f) Agencies are set up as an enterprise to provide support in financial grants and the processes, regulations, laws and review procedures that organisations need to follow. Agencies work through a number of distinct stages, beginning with the formulation of a novel idea/concept and, through a series of stages, ending in the successful launching and marketing of a new or improved product in the marketplace. Within these agencies are divisions or sections where individuals participate as technical experts in research and development, marketing, management, finance, legal, and so on, apart from outside consultants, suppliers, component manufacturers and service providers, business partners and lead users. An innovative product that meets customer expectations offers a competitive advantage. Agencies are positioned to help a company gain and retain its innovation-based advantage by diversifying with new business and experimenting with new market territories. (4 marks)

Section III

Extended-response question

Question 12

To 'analyse' this innovation is to identify components and the relationship between them and draw out and relate implications. The camera's innovation involves an improvement to the original product design with the introduction of the lens adapter and a ring light. Generally innovation occurs with the use of technologies in new and useful ways.

The key factors which support this innovation are also commonly used to gauge the innovation's success. They include the following.

- Timing: the innovation is successful when the consumer perceives a need at that point in time or companies predict the right time to launch onto the market. A company may release this camera lens innovation as an additional feature for a newly released camera.
- Available and emerging technologies: new technology or other technologies used in different ways are generally the stimulus to initiate the innovation process. The camera lens innovation is a prototype that a designer has used to develop the concept. Companies that invest in research and development (R&D) may discover ways to introduce new technologies into this product and would improve it by taking the camera to a higher end use.
- Economic and legal factors: the financial model of investing in the innovation with loans, grants and licences would need to be pursued. External factors involving laws, government agencies and regulatory bodies would need to be consulted, including patents, design registrations and copyright.
- Marketing strategies: the single biggest factor which determines the success of the camera lens innovation is its marketing plan. A variety of strategies can be brought into play including the four P's of marketing. Refer to Chapter 7 for more information.
- The role of agencies: government and allied agencies have been set up to foster the support, development and adoption of this new innovation into a productive and commercial success.

To identify the development of the camera lens as a commercialised success it is best to trace the key factors involved in its success. The success of this innovation involves its process from original idea to incubation through to commercialisation of the product. Most, if not all innovations, work through most of these elements. Refer to Chapter 15 for more information.

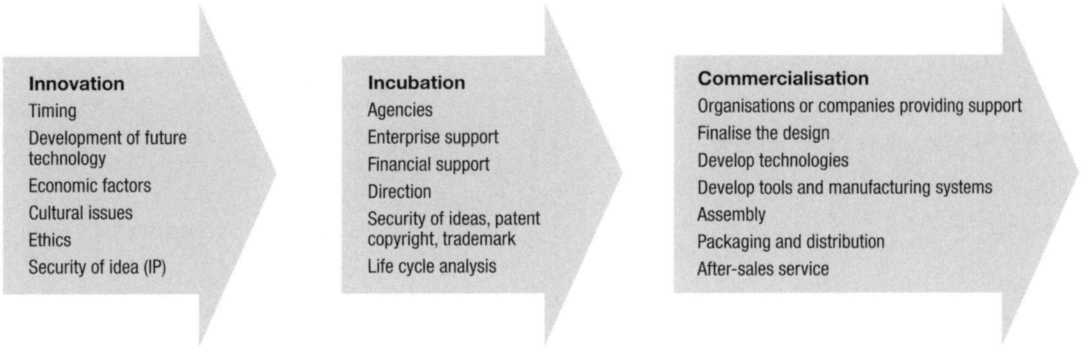

Innovation
Timing
Development of future technology
Economic factors
Cultural issues
Ethics
Security of idea (IP)

Incubation
Agencies
Enterprise support
Financial support
Direction
Security of ideas, patent copyright, trademark
Life cycle analysis

Commercialisation
Organisations or companies providing support
Finalise the design
Develop technologies
Develop tools and manufacturing systems
Assembly
Packaging and distribution
After-sales service

Marking guidelines criteria	Marks
■ Identifies the relationship between the concept of the camera's extended use and the enhancement of the final images made by this innovation ■ Identifies the relationship of the product from original idea to incubation through to commercialisation ■ Draws out and relates the implications between the ideas of the new technologies used in the camera design ■ Explains the importance of companies generally predicting the right time to be market ready ■ Identifies how R&D will help discover ways to introduce new technologies into the product that will take the product to a higher end use – Explains the importance of economic and legal factors as well as identification of marketing strategies and the role agencies have on a successful innovation – Responses well supported by returning to the stimulus image as an example	12–15
■ Identifies the concept of the camera's extended use due to the innovation ■ Identifies the concept that the success of the product is related to its ability to be commercialised ■ Draws out and relates the idea that innovation can be produced from new technologies ■ Identifies the importance of the timing of an innovation for it to be market ready ■ Identifies how R&D will help improve the product by taking the camera to a higher end use – Identifies the importance of economic and legal factors as well as identification of marketing strategies and the role agencies have on a successful innovation – Responses may be supported by returning to the stimulus image as an example	8–11
■ Identifies the camera's innovation ■ Identifies the concept that successful products need to be commercialised ■ Outlines the idea that innovation can be produced from new technologies ■ Identifies the importance of a range of strategies required to take the camera to a higher end use – Responses may be supported by returning to the stimulus image as an example	4–7
■ Recognises how the innovation can improve the use of the camera ■ Identifies the importance of some of the strategies required to take the camera to a higher end use – Responses may be supported by returning to the stimulus image as an example	3
– Recognises features of the innovation – Responses may be supported by returning to the stimulus image as an example	2
■ Recognises how a product like the camera can be changed when it is innovated	1

Sample HSC Examination 2

Section I: Objective-response questions

1. **D** is the correct option. Design is a process where the designer is given a problem to solve. The designer will use a range of approaches including researching, testing and experimentation, manufacturing and evaluation before solving the problem.
2. **D** is the correct option. Each of these steps allows the designer to move along the design continuum towards fulfilling the design brief. While the other responses are also correct, **D** covers the full range of steps.
3. **C** is the correct option. Responsible design is a personal choice about how a design will impact for future generations and on the natural world. The designer must consider environmental, social and corporate issues in order to act responsibly in an ethical sense.
4. **C** is the correct option. It is not efficient to consider every factor as only those factors that will impact on design decisions are important.
5. **A** is the correct option. The life cycle analysis considers the use of materials, their source, the energy required to obtain them and their impact on every aspect of the environment. The other responses are incorrect because they do not cover every aspect of the life cycle.
6. **D** is the correct option. Responsible design means that the designer must consider the impact that their design will have on the world. This includes the impact of resources and materials used, manufacturing and production techniques, transportation and marketing. Each of these is evaluated in terms of their environmental impact before final choices are made. The other responses only detail parts of responsible design.
7. **B** is the best option. While all factors are significant the most significant is local government regulations because unless these are adhered to the product will not be able to be manufactured.
8. **A** is the correct option. While the design being safe, using new technology and ensuring that users understand how the new technology works is important, **A** is the most significant because if the product cannot be manufactured then the design cannot be used.
9. **B** is the correct option. While it is important to ensure that new designs do not contribute to global warming and pollution is minimised; that the hole in the ozone layer is reduced; and that all materials used in production are recyclable, reusable or renewable, the aim is to keep the planet alive for future generations by lowering our carbon footprint.
10. **B** is the most correct response because evaluation is an ongoing continuous process that must occur at every stage of the design process to allow the design to move in new directions. This response also refers to both designing and creating where the other responses relate primarily to designing or creating.

Section II

Short-answer responses

Question 11

(a) Ways the design solution is visualised include discussion, thumbnail, concept and production sketches, exploded views, computerised and digital video, graphics, and models and prototypes. (2 marks)

(b) Clear communication in the client/designer relationship is essential because:

- if the communication on the design ideas is not clear the design will not be approved by the client. This can add time and cost to the work of the designer.
- it is not until the client can clearly visualise the design solution that it can be realised.

(3 marks)

(c) Intellectual property is a legal term that is used to describe who owns ideas and designs. The person who arrives at the design solution by using their own ideas owns the intellectual property. (2 marks)

(d) The designer would own the intellectual property because although the client contributed ideas, these required further consideration and needed to be worked into a possible solution via the use of the design process. (3 marks)

(e) Designers protect their intellectual property by using copyright on written design and taking out patents at each level as is required by law. For further details see Chapter 14. (2 marks)

(f) Testing and experimenting can be carried out on materials, tools, techniques and emerging technologies. This allows the designer to decide upon the best approach to take in the manufacturing process. Each test and experiment must be evaluated before conclusions are drawn. (3 marks)

Section III

Extended-response question

Question 12

The greenhouse effect or global warming: gases in the atmosphere insulate the earth, preventing the sun's heat, reflected from the earth's surface, escaping into space.

Causes include the increase of gases in the atmosphere from industrialisation and agricultural development; the burning of fossil fuels such as coal and oil which produces carbon dioxide; the increased production of chlorofluorocarbons (CFCs), nitrous oxide, hydro chlorofluorocarbons (HCFCs), and carbon dioxide from aerosols and motor vehicles.

Effects include an increase in global temperatures, climatic changes, rising sea levels and the redistribution of land suitable for agricultural production.

The ozone layer: ozone gas occurs naturally in the stratosphere 12–50 kilometres above the earth's surface. It is a protective shield from the ultraviolet radiation of the sun.

Causes include ozone is broken down naturally by UV rays but this break down is accelerated by the presence of chlorine which is found in CFCs. CFCs are produced in the manufacturing of blowing foams used in insulation and packaging materials and in aerosol sprays and in solvents for cleaning electronic components. These gases cause this ozone layer to break down, allowing dangerous UV rays to enter the atmosphere.

Effects include if the ozone layer thins or breaks more UV radiation reaches the earth's surface, causing damage to living substances. Increased UV rays cause changes to the climate and ecosystems. Currently there is a hole in the ozone layer over the North Pole which is the size of Australia and is contributing to the melting of the polar ice caps.

Tropical deforestation: a major cause for concern is the rate of destruction of the tropical rainforests. If the current rate continues it is estimated that all tropical rainforests will be destroyed in the next sixty years.

Causes include population growth causing people to cultivate forest areas; fuel required to support the population's energy requirements; commercial logging used to generate foreign currency; and consumer demand for exotic timber furniture and interiors.

Effects include the destruction of species and ecosystems containing a variety of plant and animal life; the disruption of local climates, possibly leading to desertification; increased temperatures and global climate changes due to changes in rainfall patterns; and the loss of habitat for local people. The destruction is also a significant contributing factor to the greenhouse effect.

Waste: developed countries produce over one billion tonnes of waste each year, with the average Australian household producing one tonne each. Most of this either ends up as landfill, is incinerated or is dumped in the earth or at sea.

Landfill cause: the large amounts created by the activities of humans.

Effects include that rubbish does not biodegrade into harmless substances which assimilate into the soil. Studies show that even newspaper does not break down in this airless environment. These materials often contain contaminants that leach into rivers and streams and into our drinking water. In addition, the gases released contribute to global warming unless they are tapped and used for heating. Also, no-one wants a rubbish dump near their home so waste is being transported over long distances, making the disposal both economically and environmentally costly.

Incineration cause: burning waste generates energy, although the energy needs to be close to the source or it is lost. The burning of waste causes pollution and toxic gases unless burnt at very high temperatures.

Effects include plastics and chemicals such as pesticides give off dioxins which are an extremely toxic substance. The residue that is left after incineration can contain dangerous metal pollutants which must then be collected and disposed of, often by being buried.

Dumping cause: to dispose of the vast amounts of waste humans create.

Effects include depositing rubbish beneath the earth and dumping it at sea damages both the soil and marine life. In addition, there is a limit as to how much can be dumped or buried and a real danger of poisonous contaminants leaking from damaged containers.

Water pollution cause: the growth in population and increased use of water for industrial purposes means that there is an insufficient supply of clean water to meet demand. In developing countries water is polluted from sewage, nitrites, and 'chemical cocktails' seeping from landfill sites and industrial discharges. This pollution is a problem for humans, plants and animals.

Effects include that pressure exists in industry for a 'closed loop' waste management system where no harmful chemicals can leach into water tables. However, large volumes of water are required for this so it may not be practical. The use of chlorine to bleach paper has been criticised and paper mills are instead using hydrogen peroxide which is less toxic. The use of dyes in textile mills produces harmful emissions which are not fully biodegradable.

Greenhouse effect design considerations:

- design a product, system or environment that absorbs the gases that are contributing to global warming
- design energy-efficient products that educate and inform the consumer about the greenhouse effect
- design products for recyclability rather than planned obsolescence
- design alternate power sources that do not rely on the use of fossil fuels
- improve energy conservation and efficiency in redesigning the techniques used in industry
- use efficient-energy materials to reduce the power usage in homes
- reduce dependence on cars to reduce gas emissions.

Ozone depletion design considerations:

- research alternatives to CFCs and HCFCs
- use the available alternative packaging materials
- use the available alternatives to aerosol products
- alternative insulation materials are being designed
- alternatives for refrigeration and air-conditioning gases need to be considered
- impose an environmental tax for products that are not environmentally friendly.

Water pollution design considerations:

- suppliers of manufactured products should be inspected regularly and fined if they are breaking pollution laws
- industrial processes should be checked and redesigned to meet environmental standards
- manufacturing materials need to be 'green'
- saving water should be seen as important as saving energy
- household appliances could be redesigned as water-saving devices.

Deforestation design considerations:

- planting sustainable forests
- declaring more World Heritage Areas
- more consumer education about the origin of materials and processes used in the manufacture of furnishings
- resolution of the problem of developing countries using timber as a global currency.

Waste design considerations:

1. Design a new waste removal system
2. Produce less waste by:

- increasing product life
- reducing the amount of materials used in packaging
- using biodegradable materials in the product and packaging
- reusing, recycling and remanufacturing as often as possible
- considering materials, components, planned obsolescence and replaceable components when designing a product, including designing an exterior that does not date.

References

Ekwall, W., and Shanker, J., *Diagnosis and Remediation of the Disabled Reader* (3rd ed.), Allyn & Bacon, New York, 1988.

Glasser, Dr William, *The Quality School* (revised ed.), HarperCollins, New York, 1990.

Kotler, P., *Marketing Management* (11th ed.), Prentice Hall, New York, 2002.

London, P., *No More Secondhand Art: Awakening the Artist Within*, Shambhala Publications, Boston, 1989.

Lovins, A., 'Leaner, greener world', *BWOnline* (www.businessweek.com), 23 August 2004.

Middleton, H., 'Changing Practice and Changing Lenses: The Evolution of Ways of Researching Technology Education', Centre for Learning Research.